Lecture Notes in Mathematics

Edited by A. Dold, B. Eckmann and F. Takens

1459

Dan Tiba

T0222763

Optimal Control of Nonsmooth Distributed Parameter Systems

Springer-Verlag
Berlin Heidelberg New York London
Paris Tokyo Hong Kong Barcelona

Author
Dan Tiba
Institute of Mathematics, Academy of Sciences
Bdul Păcii 220, 79622 Bucharest, Romania

Mathematics Subject Classification (1980): 49-02, 49A29, 49B22, 49B15, 49B27, 49D05

ISBN 3-540-53524-1 Springer-Verlag Berlin Heidelberg New York
ISBN 0-387-53524-1 Springer-Verlag New York Berlin Heidelberg

Printed in Germany

Printing and binding: Druckhaus Beltz, Hemsbach/Bergstr.
2146/3140-543210 – Printed on acid-free paper

INTRODUCTION

Starting with the pioneering work of Lions and Stampacchia [73], much attention is paid in the literature to the investigation of nonlinear partial differential equations involving nondifferentiable and even discontinuous terms. This includes the case of variational inequalities and free boundary problems and the main motivation of their interest is given by the many models arising in such a form with an important area of applications. See the monographs of Kinderlehrer and Stampacchia [62], Friedman [47], Ockendon and Elliott [45] where a large number of examples from various domains are discussed.

A natural direction of development of the theory is the study of related control problems and the first papers along these lines belong to Lions [71], Yvon [144], Mignot [74]. This may be also viewed as a continuation of the classical analysis of control systems governed by linear partial differential equations and we quote the wellknown book of Lions [68] in this respect.

The present work may be inscribed as a contribution to the general effort of research in nonsmooth optimization problems associated with nonlinear partial differential equations.

More precisely, the main aim of these notes is to examine distributed control problems governed by nonlinear evolution equations (parabolic or hyperbolic), in the absence of differentiability properties. In this setting, a special emphasis is given to nonlinear hyperbolic problems which are less discussed in the literature.

In order to obtain a more complete image of the area of research we have mentioned and to show other possible applications of the methods, several sections deal with problems which don't enter strictly in the announced subject: nonlinear delay differential equations, boundary control, elliptic problems and optimal design.

The material of the book is divided into three parts, after the type of the nonlinear term which occurs in the state system: semilinear and quasilinear problems, variational inequalities, free boundary problems. Among the different topics underlined throughout the work, we refer to: existence of optimal pairs, first order necessary conditions, general and efficient approximation procedures. Each time when this is possible, we indicate applications to regularity or bang-bang results for the optimal control. Examples clarifying and motivating the theory are included in every chapter.

An important role in the conception of the work is played by the methods: the adapted penalization, V.Barbu [13], the unstable systems control theory, J.L.Lions [72], the variational inequality approach [129]. From a technical point of view, in order to make these notes more useful, we have tried to use different types of arguments in different problems, of course in certain limits.

Chapter I, which introduces some fundamental notions, notations and results, is kept at a minimum length. In particular, for the existence, uniqueness and regularity theory of the various equations which appear in the text, we generally prefer to quote appropriately the scientific literature. A precious auxiliary in this respect, containing a large amount of the needed results, are the monographs of V.Barbu [12], H. Brezis [20].

The book is based mainly on the results obtained by the author during several years, but it also gives a survey of the existing literature in this area of research, via numerous comments, references, comparisons. However, as the subject is still under active development, there is no attempt to be comprehensive in any sense. We mention only the contributions of Barbu [17], Lions [72], Pawlow [90], Neittaanmaki and Haslinger [59], which are closer to our topic.

In the elaboration of this work, we received constant support and encouragement during the debates in the specialised seminars of the University of Iasi, leaded by prof.dr. V.Barbu and of the University of Bucharest, leaded by prof.dr. A.Halanay. To them and to all the other participants in the seminars, we express our gratitude. We also acknowledge with thanks the financial support of the Institute of Mathematics of the Romanian Academy of Sciences, which was decisive in the preparation of the present work.

Bucharest, June 1990 **Dan Tiba**

LIST OF SYMBOLS

R^N - the finite dimensional Euclidean space,

Ω - bounded domain in R^N,

mesΩ - the Lebesgue measure of Ω,

$\Gamma = \partial\Omega$ - the boundary of Ω,

$Q = \Omega \times]0,T[$,

$\Sigma = \partial\Omega \times [0,T]$,

$\mathcal{D}(\Omega)$, $C_0^\infty(\Omega)$ - the space of indefinitely differentiable functions, with compact support in Ω,

$\mathcal{D}'(\Omega)$ - the space of distributions on Ω,

$L^p(\Omega)$, $1 \le p \le \infty$ - the space of real functions, p-integrable in Ω (with the usual modification when $p = \infty$),

$L^p(\Omega;X)$ - the space of p-integrable functions in the sense of Bochner, with values in the Banach space X,

$C^m(\bar{\Omega};X)$ - the space of m times continuously differentiable functions on $\bar{\Omega}$, with values in X,

$W^{k,p}(\Omega)$ - the Sobolev space of real p-integrable functions, with distributional derivatives up to order k, p - integrable in Ω,

$BV(0,T;X)$ - the space of bounded variation functions on [0,T], with values in X,

$M(0,T;X)$ - the space of X-valued measures on]0,T[,

X^* - the dual of the space X,

$S(x,a)$ - the ball with centre x and radius a, in a given metric space,

$(\cdot,\cdot)_{X \times X^*}$ - the pairing between X and X^*,

$(\cdot,\cdot)_H$ - the scalar product in the Hilbert space H,

$|\cdot|_X$ - the norm in the normed space X,

$|\cdot|$ - the Euclidean norm in R^N,

dom(A) - the domain of the mapping A,

R(A) - the range of the mapping A,

A^* - the adjoint of the linear operator A,

$\mathrm{dist}(v,M)_X$ - the distance between $v \in X$ and the set $M \subset X$, in the metric of X,

int M - the interior of the set M,

$\bar{M}$ - the closure of the set M,

M x N - the cartesian product of the sets M and N,

[m,n] - an ordered pair in M x N,

cl f - the lower semicontinuous closure of f,

∂f - the subdifferential of the convex function f,

$\partial K = [-\partial_x K, \partial_y K]$ - the subdifferential of the saddle function K,

Df - the Clarke generalized gradient of f,

grad f - the gradient of the differentiable function f on R^N,

∇f - the Gateaux differential of f in normed spaces,

y', $\dot{y}$, y_t, y_x, dy/dt, $\partial y/\partial t$, $\partial y/\partial x$ - different notations of derivatives,

$\partial/\partial n$ - the outward normal derivative to Γ.

Note: sometimes we use other notations, which will be specified in the text.

CONTENTS

Pag.

INTRODUCTION III

LIST OF SYMBOLS V

Chapter I: **ELEMENTS OF NONLINEAR ANALYSIS** 1

1. Function Spaces and Compactness Principles 1
2. Monotone Operators 4
3. Generalized Gradients 11
4. Evolution Equations Associated with Monotone Operators 22

Chapter II: **SEMILINEAR EQUATION** 29

1. An Abstract Control Problem 29
2. Parabolic Problems 35
3. Hyperbolic Problems 41
4. Quasilinear Problems 60
5. Other Applications 68

Chapter III: **VARIATIONAL INEQUALITIES** 77

1. Parabolic Problems 77
2. Hyperbolic Problems 84
3. The Vibrating String with Obstacle 86
4. The Variational Inequality Method 90
5. Elliptic and Optimal Design Problems 104

Chapter IV: **FREE BOUNDARY PROBLEMS** 122

1. Two - Phase Stefan Problems 122
2. Distributed Control 128
3. Boundary Control 134
4. Discretisation 140

REFERENCES 152

I. ELEMENTS OF NONLINEAR ANALYSIS

This introductory chapter contains some prerequisites in the theory of monotone operators, convex analysis, generalized gradients, Sobolev spaces and nonlinear differential equations, collected for easier reference .

For the sake of brevity, many of the results are indicated without proof and others are omitted. The material presented here is standard and, with minor exceptions, may be found in the wellknown monographs of V.Barbu [12], [13], V.Barbu and Th.Precupanu [14], H.Brezis [20], [21], I.Ekeland and R.Temam [42], J.L.Lions [69], R.T.Rockafellar [99], D.Kinderlehrer and G.Stampacchia [62], K.Yosida [145].

1. Function spaces and compactness principles

We denote $R =]-\infty,+\infty[$, R^N, $N \in \{1,2,\ldots\}$, the finite dimensional Euclidean spaces. Let Ω be a measurable subset in R^N and X be a Banach space with norm $|\cdot|_X$. By $L^p(\Omega;X)$ we mean the space of equivalence classes, modulo the equality a.e., of functions strongly measurable in Ω, with values in X and with the norm p - integrable, $1 < p < \infty$. $L^p(\Omega;X)$ is a Banach space with respect to the norm

$$|u|^p_{L^p(\Omega;X)} = \int_\Omega |u(x)|^p_X dx .$$

For $p = \infty$, $L^\infty(\Omega;X)$ is the space of equivalence classes of functions, modulo the equality a.e., measurable from Ω to X and essentially bounded in Ω. It also is a Banach space with the norm

$$|u|_{L^\infty(\Omega;X)} = \operatorname*{ess\,sup}_{x\in\Omega} |u(x)|_X.$$

Let Ω be a bounded domain in R^N. A remarkable subspace of $L^\infty(\Omega;X)$ is $C(\bar\Omega;X)$, the Banach space of functions, continuous on $\bar\Omega$, with values in X, endowed with the topology of the uniform convergence. Two situations appear usually in the text:

- $\Omega =]0,T[\subset R$ and we use the notations $L^p(0,T;X)$, $L^\infty(0,T;X)$, $C(0,T;X)$;
- $X = R$ and we use the notations $L^p(\Omega)$, $L^\infty(\Omega)$, $C(\bar\Omega)$.

The spaces $L^p(\Omega;X)$ have wellknown properties [45], [125]. We state only the Egorov theorem, which will be of frequent use:

<u>Theorem 1.1. Assume that</u> $f_n \to f$ <u>strongly in</u> $L^p(\Omega)$, $1 \le p \le \infty$. <u>Then for every</u> $\varepsilon > 0$ <u>there is a measurable subset</u> $\Omega_\varepsilon \subset \Omega$, $mes(\Omega - \Omega_\varepsilon) < \varepsilon$ <u>and</u> $f_n \to f$ <u>uniformly on</u> Ω_ε.

We consider the space $\mathcal{D}(\Omega)$ (sometimes also denoted by $C_0^\infty(\Omega)$) of infinitely differentiable functions with compact support in Ω and its dual - the space of distributions $\mathcal{D}'(\Omega)$ to be known (see [145]). If $u \in L^p(\Omega)$, $1 \leq p \leq \infty$, then the functional

$$h \to \int_\Omega u(x)h(x)dx$$

defined on $\mathcal{D}(\Omega)$, is a distribution on Ω, denoted u, and called distribution of type function. Let k be a natural number. The Sobolev space $W^{k,p}(\Omega)$ is the space of all the distributions $u \in \mathcal{D}'(\Omega)$ of type function, such that all the distributional derivatives, $D^\alpha u$, up to order k, of u, are distributions of type function and belong to $L^p(\Omega)$. $W^{k,p}(\Omega)$ is a Banach space with the norm

$$|u|^p_{W^{k,p}(\Omega)} = \sum_{|\alpha| \leq k} \int_\Omega |D^\alpha u(x)|^p dx,$$

where α is a multiindex, $|\alpha|$ is its length and D^α is the distributional derivative of order α.

The completion of $\mathcal{D}(\Omega)$ in the topology of $W^{k,p}(\Omega)$ is denoted by $W_0^{k,p}(\Omega)$ and its dual is denoted by $W^{-k,q}(\Omega)$, $p^{-1}+q^{-1} = 1$. In the case p=2, the following notations are used $W^{k,2}(\Omega) = H^k(\Omega)$, $W_0^{k,2}(\Omega)=H_0^k(\Omega)$, $W^{-k,2}(\Omega) = H^{-k}(\Omega)$. The spaces $H^k(\Omega)$ are Hilbert spaces with the scalar product

$$(u,v)_{H^k(\Omega)} = \sum_{|\alpha| \leq k} \int_\Omega D^\alpha u . D^\alpha v dx.$$

Analogously, we define the spaces of vectorial distributions. $\mathcal{D}'(0,T;X)$ is the space of linear, continuous operators from $\mathcal{D}(]0,T[)$ to the Banach space X. $W^{k,p}(0,T;X)$ is the space of vectorial distributions $u \in \mathcal{D}'(0,T;X)$ with the distributional derivatives up to order k in $L^p(0,T;X)$. The elements of $W^{k,p}(0,T;X)$ are absolutely continuous functions together with their distributional derivatives, up to the order k - 1.

In the applications to parabolic problems, we write shortly $W^{2,1,p}(Q)$, $Q =]0,T[\times\Omega$, $p \geq 1$, for the space $L^p(0,T;W^{2,p}(\Omega)) \cap W^{1,p}(0,T;L^p(\Omega))$.

Take $\Omega = R^N$. By means of the Fourier transform

$$F : f \to Ff = (2\pi)^{-N/2} \int_{R^N} \exp(-ix.\xi)f(x)dx$$

we obtain the equivalent definition

$$H^m(R^N) = \left\{ u \in L^2(R^N); (1 + |\xi|^2)^{m/2} Fu \in L^2(R^N) \right\}.$$

In this way, one can define $H^s(R^N)$ for every $s \in R$, which remains a Hilbert space by the scalar product:

$$(u,v)_{H^s(R^N)} = ((1 + |\xi|^2)^{s/2} Fu, (1 + |\xi|^2)^{s/2} Fv)_{L^2(R^N)}.$$

If the boundary Γ of Ω is sufficiently regular, for instance if it is a C^∞ manifold, we can define the spaces $H^s(\Gamma)$ by local charts.

Proceeding by interpolation, one obtains the spaces $H^s(\Omega)$, $s \in R$. The norm in $H^s(\Omega)$

is equivalent with

$$|u|_{H^s(\Omega)} = \inf |v|_{H^s(R^N)},$$

where v is choosen such that its restriction to Ω equals u a.e. .

For the functions from $H^s(\Omega)$, $s > 0$, one may define the trace on Γ and the normal derivatives $\partial^j u/\partial n^j$ up to the order $[s-1/2]$ (the greatest positive integer majorized by $s-1/2$). This is the so called trace theorem and the proof may be found in [70].

Theorem 1.2. The mapping

$$u \to \{\partial^j u/\partial n^j;\ j = 0,1,\dots,[s-1/2]\}$$

from $\mathcal{D}(\Omega)$ to $\mathcal{D}(\Gamma)^{[s-1/2]+1}$ may be extended to a linear, continuous operator from $H^s(\Omega)$ onto $\prod H^{s-j-1/2}(\Gamma)$, $j = 0,1,\dots,[s-1/2]$.

In particular, the space $H_0^k(\Omega)$ coincides with the kernel of the trace operator.

If Ω is a bounded domain then the embedding $H^{s_1}(\Omega) \subseteq H^{s_2}(\Omega)$, $s_1 > s_2$, is compact (the Rellich theorem). Other types of embeddings are given by the Sobolev theorem:

Theorem 1.3.

i) For $q^{-1} > p^{-1} - kN^{-1}$ and $1 \le p,q \le \infty$, $k \ge 1$, then $W^{k,p}(\Omega) \subset L^q(\Omega)$. Moreover, if $p,q < \infty$, the embedding is compact.

ii) For $m < k - Np^{-1}$ we have $W^{k,p}(\Omega) \subset C^m(\bar{\Omega})$.

We continue with several other specific compactness criterions, in function spaces:

Theorem 1.4. Arzela-Ascoli

A sequence $\{x_n\}$ is relatively compact in $C(\bar{\Omega};X)$ iff:

a) it is equibounded and equicontinuous,

b) $\{x_n(z)\}$ is relatively compact in X for all $z \in \Omega$.

Theorem 1.5. Helly - Foias [68]

Let $V \subset U$ compactly, be Banach spaces. If $\{p_n\}$ is bounded in $L^\infty(0,T;V)$ and $\{(p_n)_t\}$ is bounded in $L^1(0,T;U)$ then, on a subsequence again denoted p_n, we have

$$p_n(t) \to p(t), \quad t \in [0,T]$$

strongly in U and $p \in BV(0,T;U)$.

Theorem 1.6. Dunford-Pettis

A sequence $\{f_n\}$ is weakly relatively compact in $L^1(\Omega;X)$ iff the integrals of $|f_n|_X$ are uniformly absolutely continuous on Ω.

Theorem 1.7. Lions [71]

Let B_0, B and B_1 be three Banach spaces such that $B_0 \subset B \subset B_1$ and the injection $B_0 \subset B$ is compact. Then, for every $\varepsilon > 0$ there is $C_\varepsilon > 0$ such that:

$$|u-w|_B \le \varepsilon(|u|_{B_0} + |w|_{B_0}) + C_\varepsilon |u-w|_{B_1}$$

for all $u,w \in B_0$.

Theorem 1.8. Aubin [2]

In the hypotheses of Theorem 1.7, let $\mathcal{V}$ be a bounded subset in $L^{p_0}(0,T;B_0)$ with

$$|dv/dt|_{L^{p_1}(0,T;B_1)} \leq C, \quad v \in \mathcal{V},$$

where $1 < p_0, p_1 < \infty$. Then $\mathcal{V}$ is relatively compact in $L^{p_0}(0,T;B)$.

We close this section with a variant of the Gronwall lemma, due to Brezis [20].

Proposition 1.9. Let $m \in L^1(0,T)$, $m \geq 0$ a.e. $[0,T]$ and a be a positive constant. Let $\phi \in C(0,T)$ satisfy the inequality

$$1/2\phi^2(t) \leq 1/2a^2 + \int_0^t m(s)\phi(s)ds, \quad t \in [0,T].$$

Then, we have:

$$|\phi(t)| \leq a + \int_0^t m(s)ds, \quad t \in [0,T].$$

2. Monotone operators

Consider two sets, X and Y, and X x Y their cartesian product. A subset $A \subset X \times Y$ is called a multivalued operator defined on X with values in Y. We have:

$$Ax = \{y \in Y;\ [x,y] \in A\},\ x \in X;$$
$$\mathrm{dom}(A) = \{x \in X;\ Ax \neq \phi\} \subset X;$$
$$R(A) = \bigcup_{x \in X} Ax \subset Y, \text{ the range of } A;$$
$$A^{-1} = \{[y,x];\ [x,y] \in A\} \subset Y \times X.$$

Let X be a Hilbert space. A (multivalued) operator $A : X \to X$ is called monotone if

$$(x_1 - x_2, y_1 - y_2)_X \geq 0 \quad \forall [x_i,y_i] \in A,\ i=1,2.$$

If the inequality is strict for $x_1 \neq x_2$, $x_1,x_2 \in \mathrm{dom}(A)$, then A is strictly monotone. If, moreover, the following relation holds:

$$(x_1 - x_2, y_1 - y_2)_X \geq \alpha |x_1 - x_2|_X^2, \quad \alpha > 0, \forall [x_i,y_i] \in A,\ i=1,2$$

then A is called strongly monotone.

The operator $A : X \to X$ is ω- monotone if $A + \omega I$ is monotone (the definition is useful for $\omega > 0$). The monotone operator $A : X \to X$ is called maximal monotone if its graph, as a subset in X x X, is maximal, that is it cannot be strictly included in any other monotone graph from X x X.

Proposition 2.1. Let $A : X \to X$ be a maximal monotone operator. Then:

i) A^{-1} is maximal monotone

ii) For every $x \in dom(A)$, the set Ax is convex and closed in X.

iii) A is demiclosed, that is $x_n \to x$ strongly in X, $y_n \to y$ weakly in X and $y_n \in Ax_n$ imply that $y \in Ax$.

The proof of this statement is based on the definitions and on the continuity of the scalar product. The next two results are fundamental and their proof is less elementary and we quote Barbu [12], Ch. II for a general argument.

A single valued operator $A : X \to X$ with $dom(A) = X$ is called hemicontinuous if for all $x \in X$, $y \in X$ we have $A((1-t)x + ty) \to Ax$, weakly in X, as $t \to 0$.

Proposition 2.2. A hemicontinuous monotone operator is maximal monotone.

Theorem 2.3. The following statements are equivalent:

i) A is maximal monotone in X x X.

ii) A is monotone and $R(I+A) = X$.

iii) $(I + \lambda A)^{-1}$ is a contraction (nonexpansive) mapping on the whole space X, for all $\lambda > 0$.

Let Ω be a bounded, measurable set in R^N and A be a maximal monotone operator in the Hilbert space X. It is possible to define $\tilde{A}$ on $L^2(\Omega;X)$ by $v \in \tilde{A}u$ iff $v(x) \in Au(x)$ a.e.Ω. The operator $\tilde{A}$ is maximal monotone in $L^2(\Omega;X)$. The monotonicity of $\tilde{A}$ is obvious. For the maximality, we use Thm. 2.3. The equation

$$v(x) + Av(x) \ni f(x), \quad x \in \Omega$$

with $f \in L^2(\Omega;X)$, has a unique solution $v(x) \in X$, a.e. $x \in \Omega$. As $(I + A)^{-1}$ is nonexpansive, we see that $v \in L^2(\Omega;X)$ and it is the solution of the equation $v + \tilde{A}v \ni f$.

Now, assume that Ω is a bounded domain in R^N, with smooth boundary. We consider the nonlinear differential operator:

$$Au = \sum_{|\alpha| \leq m} (-1)^{|\alpha|} D^{\alpha} A_{\alpha}(x,u, \dots ,D^m u),$$

where $A_\alpha(x,\xi)$ are real functions defined on $\Omega x R^K$, satisfying the conditions:

1) A_α are measurable in x and continuous in ξ,

2) $|A_\alpha(x,\xi)| \leq C(|\xi|^{p-1} + g(x))$

with $g \in L^q(\Omega)$, $p>1$, $q^{-1} + p^{-1} = 1$.

3) $\sum_{|\alpha| \leq m} (A_\alpha(x,\xi) - A_\alpha(x,\eta))(\xi_\alpha - \eta_\alpha) \geq 0$

for all $\xi, \eta \in R^K$ and a.e. $x \in \Omega$.

Here we denote by K the length of the vector with components u and all its derivatives up to the order m.

To the operator $A : W_0^{m,p}(\Omega) \to W^{-m,q}(\Omega)$, we associate its realization in $L^2(\Omega)$ by:

$$A_{L^2(\Omega)} u = Au, \quad u \in \mathrm{dom}(A_{L^2(\Omega)}),$$
$$\mathrm{dom}(A_{L^2(\Omega)}) = \{u \in W_0^{m,p}(\Omega);\ Au \in L^2(\Omega)\}.$$

This is a maximal monotone operator in $L^2(\Omega)$ if we assume one more condition:

$$4)(Au,u)_{W_0^{m,p}(\Omega) \times W^{-m,q}(\Omega)} \geq \alpha |u|^p_{W_0^{m,p}(\Omega)} + C, \quad \alpha > 0,$$

for all $u \in W_0^{m,p}(\Omega)$.

We recall that the operator $A : X \to X$ is coercive iff

$$\lim_{|u|_X \to \infty} (Au,u)_X / |u|_X = +\infty.$$

Theorem 2.4. A coercive, maximal monotone operator is surjective.

Proof

Let $x^* \in X$ be arbitrary fixed. By Thm 2.3, for every $\lambda > 0$, there is $x_\lambda \in X$ such that

$$(*) \qquad \lambda x_\lambda + Ax_\lambda = x^*,$$

where $A \subset X \times X$ is the given maximal monotone coercive operator. By shifting the domain of A, we may assume that $0 \in \mathrm{dom}\, A$ and multiplying (*) by x_λ, we get

$$\lambda |x_\lambda|_X \leq |A0|_X + |x^*|_X$$

by the monotonicity of A. Then, again (*), shows that $\{Ax_\lambda\}$ is bounded in X and the coercivity assumption yields that $\{x_\lambda\}$ is bounded in X. If we pass to a subsequence, we may assume that $x_\lambda \to x$ weakly in X and $Ax_\lambda \to x^*$ strongly in X. The demiclosedness of A gives $Ax = x^*$.

Remark 2.5. The monotonicity property may be defined in Banach spaces too, by replacing the inner product with the pairing $(\cdot\,,\cdot)_{X \times X^*}$. All the above properties remain true in the general setting of Banach spaces.

The nonlinear differential operator A, introduced above, is called the generalized divergence operator or the Leray-Lions operator. If we consider it as acting between the spaces $W_0^{m,p}(\Omega)$ and $W^{-m,q}(\Omega)$ we remark that it is monotone and hemicontinuous, under hypotheses 1)-3). Therefore, in this setting, it is maximal monotone without condition 4).

We say that an operator $A : X \to X$ is locally bounded in $x_0 \in X$ if there is a neighbourhood V of x_0 in X, such that $A(V) = \bigcup_{x \in V} Ax$ is a bounded subset of X.

Theorem 2.6. Any monotone operator $A : X \to X$ is locally bounded on the interior of dom(A).

Proof

By shifting dom(A), we may suppose that $0 \in$ intdom(A). If A is not locally bounded at 0, we shall derive a contradiction. Thus, consider $\{x_n\} \subset$ dom(A), $x_n^* \in Ax_n$, such that $x_n \to 0$ and $|x_n^*|_X \to \infty$ as $n \to \infty$.

We take r positive and sufficiently small such that $S(0,r) \subset$ dom(A). We show that there exist $y \in S(0,r)$ and a subsequence $n_k \to \infty$ such that

(+) $$(x_{n_k} - y, x_{n_k}^*)_X \to -\infty.$$

By contradiction, we assume that for every $u \in S(0,r)$ there is $C_u > -\infty$ such that

$$(x_n - u, x_n^*)_X \geq C_u, \quad \forall\, n \in N.$$

If $E_k = \{u \in S(0,r);\ (x_n - u,\ x_n^*)_X \geq -k,\ \forall\ n \in N\}$, then $S(0,r) = \cup E_k$ and a category argument gives that there are $\varepsilon > 0$, $k_0 \in N$, $y \in S(0,r)$ such that $S(y,\varepsilon) \subset E_{k_0}$. We get

$$(2x_n + y - u, x_n^*)_X \geq C_{-y} - k_0 \quad \forall\, n \in N,\ \forall\, u \in S(y,\varepsilon).$$

Let n_0 be such that $|x_n|_X \leq \frac{\varepsilon}{4}$ for $n \geq n_0$. Then $u = 2x_n + y - v \in S(y,\varepsilon)$ for $n \geq n_0$ and $|v|_X \leq \frac{\varepsilon}{2}$. In other words, it yields

$$(v, x_n^*)_X \geq C_{-y} - k_0,\quad \forall\, n \geq n_0,\quad \forall\, |v|_X \leq \frac{\varepsilon}{2},$$

contrary to $|x_n^*|_X \to \infty$. This shows that (+) is true and the monotonicity of A implies that

$$(x_{n_k} - y, Ay)_X \to -\infty, \text{ as } n_k \to \infty.$$

But $\{x_{n_k}\}$ is bounded and this final contradiction concludes the proof.

The next result is a generalization of Thm 2.4:

Theorem 2.7. Let A be a maximal monotone operator in X. Then A is surjective iff A^{-1} is locally bounded.

Proof

The "only if" part is a direct consequence of Thm 2.6. For the "if" part, we prove that R(A) is simultaneously a closed and open subset of X.

Let $x_0^* \in R(A) = \text{dom}(A^{-1})$ and let $\{x_n^*\} \subset R(A)$ be such that $x_n^* \to x_0^*$ as $n \to \infty$. For $x_n \in A^{-1}x_n^*$, we have $|x_n|_X \leq M$ be the hypothesis and we may assume tht $x_n \to x_0$ weakly in X.

We have

$$(x_n^* - x^*, x_n - x)_X \geq 0$$

for all $[x^*, x] \in A^{-1}$ and passing to the limit $n \to \infty$, we see that

$$(x_0^* - x^*, x_0 - x)_X \geq 0$$

for all $[x,x^*] \in A$, so $[x_0,x_0^*] \in A$ and $x_0^* \in R(A) = \overline{R(A)}$.

To show that $R(A)$ is open in X, we take $[x_0,y_0] \in A$ and $\rho > 0$ such that A^{-1} is bounded on the subset $\{ y \in R(A);\ |y - y_0|_X \leq \rho \}$. Since A is maximal monotone, the equation

$$Ax_\varepsilon + \varepsilon x_\varepsilon - \varepsilon x_0 \ni y$$

has a solution $x_\varepsilon \in \mathrm{dom}(A)$, for any $\varepsilon > 0$. We have

$$(x_\varepsilon - x_0, y_\varepsilon - y_0)_X \geq 0, \quad y_\varepsilon = y - \varepsilon x_\varepsilon + \varepsilon x_0 .$$

We take $y \in X$ with $|y - y_0|_X \leq \rho/2$ and the above inequality implies that

$$\varepsilon |x_\varepsilon - x_0|_X \leq |y - y_0|_X \leq \rho/2, \quad \forall\, \varepsilon > 0.$$

Then $|y_\varepsilon - y_0| \leq \rho$ and the boundedness properties of A yields that $\{x_\varepsilon\}$ is bounded in X. Since $y_\varepsilon \to y$ strongly in X and $x_\varepsilon \to x$ weakly on a subsequence in X, we get $[x,y] \in A$ in virtue of the maximal monotonicity of A and the set $\{ y \in X;\ |y - y_0|_X \leq \rho/2 \} \subset R(A)$ which ends the proof.

By Thm.2.3 one may define the mappings:

$$J_\lambda : X \to X, \quad J_\lambda x = (I + \lambda A)^{-1} x, \quad \lambda > 0,$$

$$A_\lambda : X \to X, \quad A_\lambda x = \lambda^{-1}(I - J_\lambda)x, \quad \lambda > 0.$$

They are called the resolvent, respectively the Yosida approximation of the maximal monotone operator A. The following properties are valid:

Theorem 2.8. Let $A : X \to X$ be maximal monotone. Then:

i) $\lim_{\lambda \to 0} J_\lambda x = \mathrm{Proj}_{\overline{\mathrm{dom}(A)}}(x), \quad x \in X;$

ii) A_λ is maximal monotone and Lipschitzian of constant $1/\lambda$ on X.

iii) For every $x \in \mathrm{dom}(A)$, we have:

$$|A_\lambda x|_X \leq |A^0 x|_X, \qquad A_\lambda x \to A^0 x, \quad \lambda \to 0$$

where $A^0 x = \mathrm{Proj}_{Ax} 0$,

iv) $A_\lambda x \in A J_\lambda x, \quad x \in X.$

Since the proof is quite lengthy we omit it and we refer to Brezis [21].

Now, we are able to state an important refinement of Proposition 2.1, iii):

Theorem 2.9. If $\lambda_n \to 0$, $x_n \to x$ weakly in X, $A_{\lambda_n} x_n \to y$ weakly in X and moreover

$$\limsup_{n,m}(x_n - x_m, A_{\lambda_n}x_n - A_{\lambda_m}x_m)_X \leq 0,$$

<u>then</u> $[x,y] \in A$ <u>and</u>

$$\lim_{n,m\to\infty}(x_n - x_m, A_{\lambda_n}x_n - A_{\lambda_m}x_m)_X = 0.$$

<u>Proof</u>

We have:

$$(x_n - x_m, A_{\lambda_n}x_n - A_{\lambda_m}x_m)_X = (J_{\lambda_n}x_n - J_{\lambda_m}x_m, AJ_{\lambda_n}x_n - AJ_{\lambda_m}x_m)_X +$$
$$+ (\lambda_n A_{\lambda_n}x_n - \lambda_m A_{\lambda_m}x_m, A_{\lambda_n}x_n - A_{\lambda_m}x_m)_X .$$

Denoting $J_{\lambda_n}x_n = \tilde{x}_n$, since $\{A_{\lambda_n}x_n\}$ is bounded in X, we get that $\tilde{x}_n \to x$ weakly in X and

$$\lim(\tilde{x}_n - \tilde{x}_m, A\tilde{x}_n - A\tilde{x}_m)_X = 0$$

by the above identity and the monotonicity of A.

We may assume that on a subsequence n_k, $(\tilde{x}_{n_k}, A\tilde{x}_{n_k})_X \to \mu$. Then

$$0 = \lim_{n_l\to\infty}[\lim_{n_k\to\infty}(\tilde{x}_{n_l} - \tilde{x}_{n_k}, A\tilde{x}_{n_l} - A\tilde{x}_{n_k})_X] = 2\mu - 2(x,y)_X$$

so $\mu = (x,y) = \lim_{n\to\infty}(x_n, Ax_n)_X$. For every $[a,b] \in A$, we have

$$(\tilde{x}_n - a, A\tilde{x}_n - b)_X \geq 0$$

and, passing to the limit, we see that

$$(x - a, y - b)_X \geq 0 \quad \forall [a,b] \in A,$$

i.e. $y \in Ax$.

Generally, it is possible that the sum of two maximal monotone operators, A + B, is not maximal monotone since, for instance, its domain may be void.

<u>Theorem 2.10. Let</u> A <u>and</u> B <u>be maximal monotone operators in</u> X x X <u>such that</u> $\mathrm{intdom}(A) \cap \mathrm{dom}(B) \neq \emptyset$. <u>Then</u> A + B <u>is maximal monotone in</u> X x X.

<u>Proof</u>

By <u>P 2.2</u> and <u>Thm 2.3</u>, the equation

$$(x) \qquad x + A_\mu x + B_\lambda x = y$$

has a unique solution denoted x_λ^μ for any $y \in X$. Here $\lambda > 0$, $\mu > 0$, A_μ and B_λ are the Yosida approximations of A, B and they are hemicontinuous. First, we keep the index μ fixed and we denote shortly x_λ, A instead of x_λ^μ, A_μ.

Without loss of generality, we may assume that $0 \in \mathrm{int}(\mathrm{dom}(A)) \cap \mathrm{dom}(B)$, $0 \in A0$, $0 \in B0$. Then (x) shows that $\{x_\lambda\}$ is bounded in X.

By Thm 2.6, there are $\rho > 0$, $M > 0$ such that

$$|x^*|_X \leq M, \quad \forall x^* \in \bigcup_{|x|_X \leq \rho} Ax .$$

We define $y_\lambda = \frac{\rho}{2}(y - B_\lambda x_\lambda - x_\lambda)/|y - B_\lambda x_\lambda - x_\lambda|_X$ and we notice that $|y_\lambda|_X \leq \frac{\rho}{2}$, so $|Ay_\lambda|_X \leq M$.

Using the monotonicity of A, we get

$$0 \leq (x_\lambda - y_\lambda, Ax_\lambda - Ay_\lambda)_X = (x_\lambda, Ax_\lambda)_X + (y_\lambda, Ay_\lambda)_X - (y_\lambda, Ax_\lambda)_X - (x_\lambda, Ay_\lambda)_X .$$

Then

$$\frac{\rho}{2}|Ax_\lambda|_X = (y_\lambda, Ax_\lambda)_X \leq \text{constant}$$

where we also use that

$$(x_\lambda, Ax_\lambda)_X = -|x_\lambda|_X^2 - (B_\lambda x_\lambda, x_\lambda)_X + (y, x_\lambda)_X \leq (y, x_\lambda)_X .$$

Thus, we have shown that $\{x_\lambda\}$, $\{B_\lambda x_\lambda\}$, $\{Ax_\lambda\}$ are bounded subsets of X and we may assume that $x_\lambda \to x_0$, $B_\lambda x_\lambda \to x_1$, $Ax_\lambda \to x_2$ weakly in X, on a subsequence. Again, by the monotonicity of A, we have

$$0 \geq (x_\lambda - x_\varepsilon, x_\lambda - x_\varepsilon)_X + (B_\lambda x_\lambda - B_\varepsilon x_\varepsilon, x_\lambda - x_\varepsilon)_X ,$$

therefore

$$\limsup_{\lambda,\varepsilon \to 0} (B_\lambda x_\lambda - B_\varepsilon x_\varepsilon, x_\lambda - x_\varepsilon)_X \leq 0$$

and Thm 2.9 gives that $x_1 \in Bx_0$ and

$$\lim_{\lambda,\varepsilon \to 0} (B_\lambda x_\lambda - B_\varepsilon x_\varepsilon, x_\lambda - x_\varepsilon)_X = 0 .$$

Consequently

$$\limsup_{\lambda,\varepsilon \to 0} (x_\lambda - x_\varepsilon, x_\lambda + Ax_\lambda - x_\varepsilon - Ax_\varepsilon)_X = 0$$

and applying once more Thm 2.9 to the operator I + A, we see that $x_2 = A_\mu x_0$ (we recall that in fact A stands for A_μ up to now).

Assume that $0 \in \mathrm{intdom}(B)$, for the moment. The same argument as above allows to take $\mu \to 0$. Then, we see that the equation

$$x_\lambda + Ax_\lambda + B_\lambda x_\lambda \ni y$$

has a unique solution under the conditions of Thm 2.10. Since $0 \in \operatorname{intdom}(A)$, we can iterate the above proof and we obtain the desired conclusion by Thm 2.3.

Remark 2.11. There is a strong relationship between the maximal monotone operators and the nonlinear contraction semigroups. This idea will be stressed in the last section.

3. Generalized gradients

3.1. The subdifferential of a convex function

A remarkable class of monotone operators is given by the subdifferentials of convex functions.

Let X be a Banach space and $\varphi: X \to [-\infty, +\infty]$ be a convex function. Then, we define $\operatorname{dom}(\varphi) = \{x \in X; \varphi(x) < +\infty\}$ and φ is called proper if $\operatorname{dom}(\varphi) \neq \emptyset$ and $\varphi(x) > -\infty$ for all $x \in X$.

The closure of φ, denoted $\operatorname{cl}\varphi$, is the lower semicontinuous hull of φ:

$$(\operatorname{cl}\varphi)(x) = \liminf_{y \to x} \varphi(y),$$

if $\varphi(x) > -$ for all $x \in X$, and $\operatorname{cl}\varphi = -\infty$ otherwise. The convex function φ is said to be closed if $\varphi = \operatorname{cl}\varphi$. In particular, for a proper, convex function, the closedness is equivalent with the lower semicontinuity.

The subdifferential of the function φ, denoted $\partial\varphi$, is the (possibly multivalued) operator in $X \times X^*$, given by

$$\partial\varphi(x) = \left\{ w \in X^*; \varphi(x) - \varphi(v) \leq (w, x-v)_{X \times X^*}, \forall v \in X \right\} .$$

Obviously, x_0 is a minimum point for φ iff $0 \in \partial\varphi(x_0)$.

Theorem 3.1. In Banach spaces, the subdifferential of a lower semicontinuous, proper, convex function φ is a maximal monotone operator.

Proof

We give the argument for the case of reflexive Banach space X. For the general situation, we quote Rockafellar [98]. By the renorming theorem of Asplund [1], we may assume that X and X^* are strictly convex too.

The monotonicity of $\partial\varphi$ is obvious by the definition.

According to the extension of Thm 2.3 to Banach spaces, we have to show that the equation

$$Fx + \partial\varphi(x) \ni x^*$$

has a solution for any $x^* \in X^*$. Here $F : X \to X^*$ is the duality mapping (see Example 1 at the end of this paragraph).

Let $f : X \to R$ be the convex, lower semicontinuous function given by

$$f(x) = \frac{1}{2}|x|_X^2 + \varphi(x) - (x,x^*)_{X \times X^*}.$$

By Thm 3.2–3.4, which will be proved independently, we notice that $\lim_{|x|_X \to \infty} f(x) = \infty$ and there is $x_0 \in X$ such that

$$f(x_0) \leq f(x), \quad \forall x \in X.$$

We rewrite this inequality in the form

$$\varphi(x_0) - \varphi(x) \leq (x_0 - x,x^*)_{X \times X^*} + (x - x_0,Fx)_{X \times X^*}, \quad \forall x \in X$$

and take $x = tx_0 + (1 - t)u$, $t \in [0,1]$, $u \in X$. By the convexity of φ, we infer:

$$\varphi(x_0) - \varphi(u) \leq (x_0 - u,x^*)_{X \times X^*} + (u - x_0,F(tx_0 + (1 - t)u))_{X \times X^*}$$

Passing to the limit $t \to 1$ and using the demicontinuity of F, we get

$$\varphi(x_0) - \varphi(u) \leq (x_0 - u,x^*)_{X \times X^*} + (u - x_0,F(x_0))_{X \times X^*},$$

i.e. $x^* - F(x_0) \in \partial\varphi(x_0)$ by the definition of $\partial\varphi$.

We list several properties of convex functions, which play an important role in optimization:

Theorem 3.2. Any proper, closed, convex function φ attains its minimum value on closed, convex bounded subsets of a reflexive Banach space X.

Proof

A bounded, closed convex subset of X is weakly compact and φ is weakly lower semicontinuous according to Thm 3.3. We conclude the argument by the Weierstrass theorem.

If φ is coercive in the sense that

$$\lim_{|x|_X \to \infty} \varphi(x) = +\infty$$

then the boundedness assumption in Thm.3.2 is no longer necessary.

Theorem 3.3. A proper, convex function is lower semicontinuous iff it is weakly lower semicontinuous.

Proof

This is a consequence of the Mazur theorem (Yosida [145]) applied to the epigraph of the convex function φ

$$\text{epi}\,\varphi = \{[x,\alpha];\ x\in X,\ \alpha\in R,\ \varphi(x)\le\alpha\}.$$

Proposition 3.4. Let $\varphi: X\to]-\infty,+\infty]$ be a convex, closed proper function. Then:
i) φ is bounded from below by an affine function;

ii) $\text{dom}(\partial\varphi)\subset\text{dom}(\varphi)$, $\overline{\text{dom}(\partial\varphi)} = \overline{\text{dom}(\varphi)}$, $\text{intdom}(\varphi) = \text{intdom}(\partial\varphi)$.

Proof

We prove only i), which is used in the proof of Thm 3.1. For ii), which needs a more complex argument, we quote the book of Barbu and Precupanu [14], Ch. II. As φ is proper, lower semicontinuous, convex, then epiφ is a nonvoid, closed convex subset of $X \times R$. If $x_0\in\text{dom}(\varphi)$, then $[x_0,\varphi(x_0)-\varepsilon]\notin\text{epi}\,\varphi$, $\forall\varepsilon>0$. The Hahn-Banach theorem yields the existence of $x_0^*\in X^*$ and $\alpha\in R$ such that

$$\sup_{[x,t]\in\text{epi}\varphi}\left\{x_0^*(x)+t\alpha\right\} < x_0^*(x_0)+\alpha(\varphi(x_0)-\varepsilon)$$

(we have identified $(X\times R)^*$ with $X^*\times R$).

Obviously $\alpha<0$ since $[x_0,\varphi(x_0)+n]\in\text{epi}\varphi$, $\forall n\in\mathbf{N}$. But $[x,\varphi(x)]\in\text{epi}\varphi$, $\forall x\in\text{dom}(\varphi)$, thus

$$x_0^*(x)+\alpha\varphi(x)\le x_0^*(x_0)+\alpha\varphi(x_0),\quad \forall\ x\in\text{dom}(\varphi).$$

Equivalently, we get

$$\varphi(x)\ge-\frac{1}{\alpha}x_0^*(x)+\frac{1}{\alpha}x_0^*(x_0)+\varphi(x_0)$$

and the proof is finished.

For any function $f: X\to\overline{R}=[-\infty,+\infty]$, we define the Fenchel conjugate $f^*: X^*\to\overline{R}$ by

$$f^*(x^*)=\sup\{(x,x^*)_{X\times X^*}-f(x);\ x\in X\},\quad x^*\in X^*.$$

Similarly, one may define the biconjugate $f^{**}: X\to\overline{R}$ (with respect to the pairing between X and X^*) and the conjugate of order n, denoted $f^{(n)*}$. As easy consequences of the definition we have the so called Young inequality

$$(x,x^*)_{X\times X^*}\le f(x)+f^*(x^*),\quad\forall x\in X,\ \forall x^*\in X^*,$$

the inequality $f^{**}\le f$ and the property that f^* and f^{**} are always convex and lower semicontinuous in the weak* topology of X^* and in the weak topology of X, respectively. The mapping f^{**} is in fact the Γ-regularization of f (Ekeland and Temam [42]) and $f^{**}=f$ if f is convex, lower semicontinuous, proper.

By the Young inequality, we notice that, if f^* is proper, then Thm 3.3 i) follows. The converse is also valid (from the definition) for lower semicontinuous, convex, proper functions f.

As another important consequence of the Young inequality in reflexive Banach spaces, we have $\partial f^* = (\partial f)^{-1}$, when f is a convex, lower semicontinuous proper function.

Proposition 3.5. A convex, lower semicontinuous, proper function is locally Lipschitzian on the interior of its domain.

This is an immediate consequence of the definition of the subdifferential and of the local boundedness property of maximal monotone operators.

For $\lambda > 0$, we define the (Yosida) regularization φ_λ of φ by:

$$\varphi_\lambda(x) = \inf\left\{|x-y|_X^2/2\lambda + \varphi(y); y \in X\right\}, \quad x \in X.$$

Theorem 3.6. The function φ_λ is convex, everywhere finite on X. If X is a Hilbert space, then φ_λ is Frechet differentiable on X. Moreover, denoting $A = \partial\varphi$, then $A_\lambda = \partial\varphi_\lambda$. We have:

i) $\varphi(J_\lambda x) \leq \varphi_\lambda(x) \leq \varphi(x), \quad x \in X, \ \lambda > 0;$

ii) $\lim_{\lambda \to 0} \varphi_\lambda(x) = \varphi(x), \quad x \in X.$

Here $J_\lambda = (I + \lambda A)^{-1} = (I + \lambda\partial\varphi)^{-1}$.

Proof

The subdifferential of the mapping

$$y \to |x - y|_X^2/2\lambda + \varphi(y), \quad y \in X$$

is just $\lambda^{-1}(y - x) + \partial\varphi(y)$. Thus, the infimum defining φ_λ is attained a point x_λ such that

$$x_\lambda - x + \lambda\partial\varphi(x_\lambda) \ni 0,$$

that is $x_\lambda = J_\lambda x$ and we obtain i) by the definition of φ_λ.

If $x \in \mathrm{dom}(\varphi)$, then $\lim_{\lambda \to 0} J_\lambda x = x$ and ii) follows by i) and the lower semicontinuity of φ. If $x \notin \mathrm{dom}(\varphi)$, then $\varphi(x) = +\infty$. We assume (by contradiction) that there is a subsequence λ_n such that $\lambda_n \to 0$ and $\varphi_{\lambda_n}(x) \leq C$. We get

$$|J_{\lambda_n}(x) - x|_X^2/2\lambda_n + \varphi(J_{\lambda_n} x) \leq C,$$

so $J_{\lambda_n}(x) \to x$ strongly in **X** and the lower semicontinuity of φ gives $\varphi(x) \leq C$ contradicting $\varphi(x) = +\infty$.

To finish the proof, we remark the following inequality

$$\varphi_\lambda(y) - \varphi_\lambda(x) \leq \frac{\lambda}{2}(|A_\lambda y|_X^2 - |A_\lambda x|_X^2) + (A_\lambda y, J_\lambda y - J_\lambda x)_X$$

where $A_\lambda = (\partial\varphi)_\lambda = (I - J_\lambda)/\lambda$.

On the other side, we have

$$\varphi(J_\lambda y) - \varphi(J_\lambda x) \geq (A_\lambda x, J_\lambda y - J_\lambda x)_X$$

that is

$$\varphi_\lambda(y) - \varphi_\lambda(x) \geq \frac{\lambda}{2}[\,|A_\lambda y|_X^2 - |A_\lambda x|_X^2] + \lambda(A_\lambda x - A_\lambda y, A_\lambda x)_X + (A_\lambda x, y - x)_X \ .$$

Then, we get

$$\varphi_\lambda(y) - \varphi_\lambda(x) - (A_\lambda x, y - x)_X \geq \frac{\lambda}{2}|A_\lambda y - A_\lambda x|_X^2 \geq 0.$$

Combining the above inequalities, we infer:

$$0 \leq \varphi_\lambda(y) - \varphi_\lambda(x) - (A_\lambda x, y - x)_X \leq (A_\lambda y - A_\lambda x, y - x)_X \ .$$

Since A is Lipschitz with constant $1/\lambda$, we obtain the Frechet differentiability of φ_λ and the relation $A_\lambda = \partial\varphi_\lambda$.

We continue with a result on the subdifferential of a convex function composed with a linear operator, following [112]. See also [140] for an extension to more general spaces.

Let X be a reflexive Banach space and $\varphi: X \to]-\infty, +\infty]$ be a closed, proper, convex function. Let $A : X \to X$ be a linear, bounded operator, with adjoint A^*. We consider the composed function $\psi: X \to]-\infty, +\infty]$:

$$\psi(x) = \varphi(Ax), \quad x \in X.$$

If $R(A) \cap \mathrm{dom}(\varphi) \neq \emptyset$, then ψ is convex, proper, lower semicontinuous.

Theorem 3.7. Assume that X may be decomposed into the direct sum $X = X_1 \oplus X_2$, such that:

i) $R(A) \cap \mathrm{int}_1[\mathrm{dom}(\varphi) \cap X_1] \neq \emptyset$,

ii) $A^*|_{X_1^*} : X_1^* \to X^*$ has bounded inverse.

Then, we have the formula $\partial\psi(x) = A^*\partial\varphi(Ax)$, $x \in X$.

Here int_1 denotes the interior in the relative topology of X_1.

Proof (sketch)

By the definition of the subdifferential, we have

$$A^*\partial\varphi(Ax) \subset \partial\psi(x), \qquad x \in X.$$

Let $S : X \to X^*$ be the (multivalued) operator $Sx = A^*\partial\varphi(Ax)$. S is monotone and we show that it is maximal. According to the extension of Thm.2.3 to Banach spaces [12], this is equivalent with $R(F + S) = X^*$, where F is the duality mapping in X (see Example 1 at the end of this paragraph).

For $f \in X^*$, we consider the approximating equation

(*) $$Fx_\lambda + A^*\partial\varphi_\lambda(Ax_\lambda) = f, \ \lambda > 0.$$

This has a solution $x_\lambda \in X$ since $A^*\partial\varphi_\lambda A$ is monotone and hemicontinuous, that is maximal monotone.

Assume that $0 \in R(A) \cap \mathrm{int}_1[\mathrm{dom}(\varphi) \cap X_1]$. The general situation may be discussed similarly. We multiply (*) by x_λ and we use the definition of the subdifferential and Proposition 3.4 to deduce that $\{x_\lambda\}$ is bounded in X.

Now, we write

$$\partial\varphi_\lambda(Ax_\lambda) = \partial_1\varphi_\lambda(Ax_\lambda) + \partial_2\varphi_\lambda(Ax_\lambda), \ \partial_1\varphi_\lambda(Ax_\lambda) \in X_1^*, \ \partial_2\varphi_\lambda(Ax_\lambda) \in X_2^*,$$

where X_1^*, X_2^* are the polar subspaces of X_1 and X_2 and $X^* = X_1^* \oplus X_2^*$. Then

$$Fx_\lambda + A^*\partial_1\varphi_\lambda(Ax_\lambda) + A^*\partial_2\varphi_\lambda(Ax_\lambda) = f.$$

By assumption i) and Proposition 3.5 we can find $\bar{\rho} > 0$ sufficiently small such that

$$\varphi(\rho h) \le C, \quad \rho \in [-\bar{\rho}, \bar{\rho}], \quad h \in X_1, \ |h|_X = 1.$$

We have:

$$(\partial_2\varphi_\lambda(Ax_\lambda), \rho h) \le (\partial\varphi_\lambda(Ax_\lambda), Ax_\lambda) + \varphi_\lambda(\rho h) - \varphi_\lambda(Ax_\lambda) \le$$

$$\le (f, x_\lambda) - |x_\lambda|_X^2 + \varphi(\rho h) - \varphi_\lambda(Ax_\lambda) \le C$$

for $|\rho| \le \bar{\rho}$, $h \in X_1$, $|h|_X = 1$.

Here C denotes several constants and $(\cdot\,,\cdot)$ is the pairing between X and X^*.

It yields that $\{\partial_2\varphi_\lambda(Ax_\lambda)\}$ is bounded in X_2^*, that is in X^*. By (*) and hypothesis ii) we get $\{\partial\varphi_\lambda(Ax_\lambda)\}$ bounded in X^*.

To finish the proof, one may pass to the limit by means of Thm.2.9.

The perturbation result given by Thm 2.10 becomes more precise in the case of subdifferential mappings:

Theorem 3.8. Let $A : X \to X$ be a maximal monotone operator and $\varphi : X \to]-\infty, +\infty]$ be a convex, lower semicontinous, proper function. We assume one of the following two conditions:

i) $\mathrm{dom}(A) \cap \mathrm{int}\,\mathrm{dom}(\varphi) \ne \emptyset$,

ii) $\mathrm{dom}(\varphi) \cap \mathrm{int}\,\mathrm{dom}(A) \ne \emptyset$.

Then $A + \partial\varphi$ is a maximal monotone operator.

For the proof and another generalization and unification of Thms.3.7, 3.8 we refer to Aubin [3], Ch.8.

We close this paragraph with some examples of subdifferential operators, important in the sequel.

Example 1

The (multivalued) operator $F : X \to X^*$, defined by

$$Fx = \{x^* \in X^*; (x,x^*)_{X \times X^*} = |x|_X^2 = |x^*|_{X^*}^2\}$$

is called the duality mapping of X. It coincides with the subdifferential of the convex function $f : X \to R$, $f(x) = 1/2|x|_X^2$ as it may be easily checked by the definition.

We remark that F is homogeneous in Banach spaces and it is additive only in Hilbert spaces when it coincides with the Riesz isomorphism. Moreover, F is surjective iff X is reflexive and strictly monotone iff X is strictly convex (Barbu and Precupanu [14], Ch. I). In reflexive spaces with strictly convex dual, F is single valued and demicontinuous.

Example 2

Any maximal monotone graph β in RxR is ciclically monotone and there is a convex, lower semicontinous, proper function $j : R \to]-\infty,+\infty]$ such that $\beta = \partial j$. If β^0 is the principal section of β ($|\beta^0(x)| = \min|\beta(x)|$) and $]a,b[$ is an interval such that $]a,b[\subset \mathrm{dom}\,\beta \subset \mathrm{dom}\, j \subset [a,b]$ (a and b may be infinite), then β^0 is a nondecreasing function on $]a,b[$ and $\beta(x) = [\beta^0(x-),\beta^0(x+)]$ for all $x \in]a,b[$. Moreover, if $a \in \mathrm{dom}\,\beta$ (respectively $b \in \mathrm{dom}\,\beta$), then $\beta(a) =]-\infty, \beta^0(a+)]$ (respectively $\beta(b) = [\beta^0(b-), +\infty[$).

Example 3

Let $g : R \to]-\infty,+\infty]$ be a convex, lower semicontinous, proper function and $\varphi : L^2(\Omega) \to$ $\to]-\infty,+\infty]$ be given by:

$$\varphi(u) = \begin{cases} \int_\Omega g(u(x))dx & \text{if } g(u) \in L^1(\Omega) \\ +\infty & \text{otherwise.} \end{cases}$$

The mapping φ is convex, lower semicontinuous, proper on $L^2(\Omega)$ and $w \in \partial\varphi(u)$ iff $w(x) \in \partial g(u(x))$ a.e.Ω, $w \in L^2(\Omega)$.

Example 4

We consider the convex, closed, proper function φ defined on $L^2(\Omega)$ by

$$\varphi(u) = \begin{cases} 1/2 \int_\Omega |\mathrm{grad}\, u|^2 dx + \int_\Omega g(u(x))dx, & \text{if } u \in H_0^1(\Omega), g(u) \in L^1(\Omega), \\ +\infty & \text{otherwise.} \end{cases}$$

Then

$$\mathrm{dom}(\partial\varphi) = \{u \in H_0^1(\Omega) \cap H^2(\Omega);\ u(x) \in \mathrm{dom}(\partial g) \text{ a.e.}\Omega\},$$

$$\partial\varphi(u) = \{w \in L^2(\Omega);\ w(x) \in -\Delta u(x) + \partial g(u(x)) \text{ a.e.}\Omega\},$$

according to Barbu [12], p.203.

3.2. The subdifferential of a saddle function

Let X and Y be two Hilbert spaces and $L : X \times Y \to]-\infty,+\infty]$ be a convex, lower semicontinous, proper function. The operator ∂L has values in XxY and may be written in the form

$$\partial L(x,y) = [\partial_1 L(x,y), \partial_2 L(x,y)],$$

where $\partial_1 L$, $\partial_2 L$ signify the first, respectively the second component of the ordered pair. They don't coincide with the partial subddiferentials, except the case when L is Frechet differentiable.

We call Hamiltonian associated with the convex function L, its partial Fenchel conjugate, $K : X \times Y \to [-\infty,+\infty]$, given by

$$K(x,y) = \sup\left\{(y,z)_Y - L(x,z);\ z \in Y\right\}.$$

The function K is concave-convex on X x Y. Generally, for a concave - convex (we also say "saddle") function, we denote:

$$\mathrm{dom}_1(K) = \left\{x \in X;\ cl_2 K(x,y) > -\infty, \text{ for all } y \in Y\right\},$$
$$\mathrm{dom}_2(K) = \left\{y \in Y; cl_1 K(x,y) < +\infty, \text{ for all } x \in X\right\}.$$

Here $cl_2 K$ is obtained by closing the mapping $K(x,.)$ as a convex function on the second argument, and $cl_1 K$ is obtained similarly by closing $K(.,y)$ as a concave function.

Two saddle functions K and M are called equivalent if $cl_i K = cl_i M$, $i = 1,2$. The saddle function K is closed if $cl_1 K$ and $cl_2 K$ are equivalent with K. It is proper if

$$\mathrm{dom}(K) = \mathrm{dom}_1(K) \times \mathrm{dom}_2(K)$$

is nonvoid.In Banach spaces, the subdifferential of K is the (multivalued) operator $\partial K : X \times Y \to X^* \times Y^*$, given by

$$\partial K(x,y) = [-\partial_x K(x,y), \partial_y K(x,y)],$$
$$\partial_y K(x,y) = \left\{y^* \in Y^*; K(x,y) \leq K(x,v) + (y - v, y^*)_{Y \times Y^*}\ v \in Y\right\},$$
$$\partial_x K(x,y) = \left\{x^* \in X^*; K(u,y) \leq K(x,y) + (u - x, x^*)_{X \times X^*},\ u \in X\right\}.$$

It is easy to see that $[x_o,y_o]$ is a saddle point for K iff $[0,0] \in \partial K(x_o,y_o)$.

Theorem 3.9. Let Y be a reflexive Banach space and $K : X \times Y \to [-\infty,+\infty]$ be a closed, proper, saddle function. Then, the mapping $\partial K : X \times Y \to X^* \times Y^*$ is maximal monotone.

Proof (sketch)

The monotonicity of ∂K is obvious by the definition.

We associate to K its partial conjugate $L : X \times Y^* \to R$ by

$$L(x,y^*) = \sup\left\{ (y,y^*)_{Y\times Y^*} - \quad (x,y);\ y\in Y \right\},$$

which is convex lower semicontinuous. By the properties of conjugate functions, it turns out that K is the Hamiltonian associated to L. The relationship between the subdifferentials of conjugate mappings, in reflexive spaces, gives that:

$$[x^*,y^*] \in \partial K(x,y)$$

iff

$$[x^*,y] \in \partial L(x,y^*).$$

Thenthe maximality of ∂K follows by the maximality of ∂L.

For a detailed argument along these lines, we quote Barbu and Precupanu [14].

All the results of §2 apply to ∂K. In particular, the Yosida approximation $(\partial K)_\lambda$ may be precised as in Thm.3.6, by defining the regularization of K, according to [96]:

$$K_\lambda^\lambda(x,y) = \sup_{u\in X} \inf_{v\in Y} \left\{ -|x-u|_X^2/2\lambda + |y-v|_Y^2/2\lambda + K(u,v) \right\}.$$

Theorem 3.10. *Let K be a closed, proper, concave - convex function on X x Y and $A = \partial K$. The mapping K_λ^λ is Gateaux differentiable on X x Y and $A_\lambda = \partial K_\lambda^\lambda$. Moreover*

$$cl_2 K \le \underline{\lim}_{\lambda\to 0} K_\lambda^\lambda \le \overline{\lim}_{\lambda\to 0} K_\lambda^\lambda \le cl_1 K.$$

Remark 3.11. Another type of maximal monotone operator, associated with a saddle function, was introduced by E.Krauss [63]. Thm 3.10 also extends to this setting [126]. It is also possible to define "anisotropic regularizations", K_δ^λ, by the same formula, [113], [127], and to prove similar results.

3.3. The Clarke generalized gradient

In many situations, optimization problems with nondifferentiable and nonconvex functionals appear. To obtain first order necessary conditions, it is useful to have a notion to replace the classical derivative or the subdifferential. A general definition along these lines was introduced by F.H.Clarke [31]. See also his monograph [32] for a detailed analysis of this and of related concepts and their applications in the calculus of variations and optimal control.

Let $f : R^N \to R$ be locally Lipschitzian. The Clarke generalized gradient of f in $x\in R^N$ is denoted $Df(x)$ and it is the convex hull of the set of cluster points for the sequences grad $f(x + h_i)$, where $h_i \to 0$ are choosen such that the gradient exists:

$$Df(x) = \operatorname{conv}\left\{ w\in R^N; \exists\, h_i\to 0, \exists\, \mathrm{grad}(x+h_i)\to w \right\}.$$

The next two propositions are direct consequences of the definition of Df and we omit the proofs.

Proposition 3.12. The generalized gradient of a locally Lipschitzian mapping is a multivalued operator with nonvoid, compact, convex values and it is upper semicontinuous in the sense of multifunctions:

$$v_i \to v,\ x_i \to x,\ v_i \in Df(x_i) \Rightarrow v \in Df(x).$$

This notion may be extended to the class of semicontinuous real functions, but, in the applications, we shall limit ourselves to the locally Lipschitzian case.

If f is convex, then $Df = \partial f$. Moreover, we have the characterization.

Proposition 3.13. The following statements are equivalent:

i) $Df(x) = \{\xi\}$, a set with one element;

ii) There exists grad $f(x) = \xi$ and grad f is continuous relatively to the subset on which it exists.

The following "demiclosedness" property of the Clarke generalized gradient is useful in applications and extends the upper semicontinuity of Df.

Theorem 3.14. Let E be a locally compact space with a positive measure ν, $\nu(E) < +\infty$. Let $\{y_\varepsilon\} \subset L^1(E)$ be a sequence satisfying, for $\varepsilon \to 0$:

$$y_\varepsilon \to y \ \text{strongly in } L^1(E),$$
$$\dot{\beta}^\varepsilon(y_\varepsilon) \to g \ \text{weakly in } L^1(E).$$

Then $g(x) \in D\beta(y(x))$ ν-a.e. in E.

Here β: $R \to R$ is locally Lipschitzian and β^ε is a smoothing of β

$$\beta^\varepsilon(y) = \int_{-\infty}^{\infty} \beta(y - \varepsilon s)\rho(s)ds,$$

with $\rho \in C_o^\infty(R)$, supp$\rho \subset [-1,1]$, $\rho(-s) = \rho(s)$, $\rho(s) \geq 0$, $\int_{-\infty}^{\infty} \rho(s)ds = 1$.

Proof (Barbu [15])

On a subsequence, we may assume that

$$y_\varepsilon(x) \to y(x) \quad \nu\text{-a.e. } x \in E.$$

On the other side, the Mazur theorem, Yosida [145], gives that

$$g = \lim_{m \to \infty} g_m \quad \text{strongly in } L^1(E),$$

where $\{g_m\} \subset L^1(E)$ are of the form

$$g_m = \sum_{j \in I_m} \alpha_m^j \dot{\beta}^{\varepsilon_j}(y_{\varepsilon_j})$$

with I_m a finite subset of natural numbers from the interval $[m,\infty[$ and $\alpha^j_m \geq 0$, $\sum_{j\in I_M} \alpha^j_m = 1$.

Then, we may also assume that

$$g_m(x) \to g(x) \quad \nu\text{-a.e. } x \in E.$$

We fix $x \in E$ satisfying both properties and consider $\{z_n\} \subset R$ such that $\dot\beta(z_n)$ exists and $z_n \to y(x)$. We have

$$\dot\beta^{\varepsilon_j}(y_{\varepsilon_j}) = \varepsilon_j^{-1} \int_{-\infty}^{\infty} \beta(y_{\varepsilon_j} - \varepsilon_j\theta)\rho'(\theta)d\theta \ .$$

Developing $\beta(y_{\varepsilon_j} - \varepsilon_j\theta)$ and $\beta(y_{\varepsilon_j})$ around z_j, we have

$$(\dot\beta^{\varepsilon_j})(y_{\varepsilon_j}) - \dot\beta(z_j) = \varepsilon_j^{-1} \int_{-\infty}^{\infty} \omega_j(y_{\varepsilon_j} - z_j)\,\rho'(\theta)d\theta -$$

$$- \varepsilon_j^{-1} \int_{-\infty}^{\infty} w_j(z_j - y_{\varepsilon_j} + \varepsilon_j\theta)\,\rho'(\theta)d\theta$$

with ω_j, $w_j \to 0$ uniformly in θ .

On the other hand z_j can be chosen sufficiently close to y_{ε_j} such that $|y_{\varepsilon_j} - z_j|/\varepsilon_j \to 0$ for $j \to \infty$. Thus, we have

$$|\dot\beta^{\varepsilon_j}(y_{\varepsilon_j}) - \dot\beta(z_j)| \to 0 \quad \text{for } j \to \infty,$$

which, along with the definition of $D\beta$, yields $g(x) \in D\beta(y(x))$.

Remark 3.15. The statement repeats identically when β is monotone and locally Lipschitzian and we have

$$\beta^{\varepsilon}(y) = \int_{-\infty}^{\infty} \beta_\varepsilon(y - \varepsilon s)\rho(s)ds,$$

where β_ε is the Yosida approximation of β.

Remark 3.16. For a detailed treatment of the theory of monotone operators and convex functions, in connection with their differentiability properties and with applications in optimization theory, we mention the recent monograph of R. Phelps [97].

We close this section with a variant of the so called Ekeland's variational principle [43], [44], which may be viewed as an important extension of the classical Fermat theorem and is of frequent use in the recent literature on control theory.

Theorem 3.16. Let (E,d) be a complete metric space and F a l.s.c. mapping: $E \to]-\infty,+\infty]$, bounded from below. For any $\varepsilon > 0$, let $e_\varepsilon \in E$ be such that $F(e_\varepsilon) \leq \inf F + \varepsilon^2$. Then, there exists $e' \in E$ satisfying

$$F(e') \leq F(e_\varepsilon),$$

$d(e_\varepsilon, e') \leq \varepsilon$,

$F(e') \leq F(e) + \varepsilon\, d(e,e'), \quad \forall\, e \in E.$

4. Evolution equations associated with monotone operators

Let X be a Banach space and $C \subset X$ be a closed subset. A contraction semigroup on C is a mapping $S : [0, +\infty[\times C \to C$, which satisfies

i) $S(t + s)x = S(t)S(s)x, \quad x \in C,\ t,s \geq 0,$

ii) $S(0)x = x, \qquad x \in C,$

iii) for all $x \in C$, $S(t)x$ is continuous with respect to $t \geq 0$,

iv) $|S(t)x - S(t)y|_X \leq |x - y|_X, \quad t \geq 0,\ x,y \in C.$

If S is a contraction semigroup on C, the infinitesimal generator A of S, is defined by

$$Ax = \lim_{h \to 0} (S(h)x - x)/h, \qquad x \in dom(A),$$

$$dom(A) = \{x \in C;\ \text{the limit exists in } X\}.$$

In Hilbert spaces, a complete theory of the generation of contraction semigroups, may be obtained

Theorem 4.1. Let C be a closed, convex, nonvoid set in X, a Hilbert space, and $\{S(t),\ t \geq 0\}$ be a contraction semigroup on C. Then, there is a unique maximal monotone operator $A : X \to X$ such that $-A^0$ is the generator of S.

Conversely, if A is a maximal monotone subset in $X \times X$, there is a unique semigroup S on dom(A) such that $-A^0$ is the generator of S.

Here A^0 is, as usual, the minimal section of A.

This may be viewed as a generalization of the Hille-Yosida-Phillips characterization of generators of linear semigroups. For the proof we quote the books of Barbu [12] and Brezis [21].

Let A be a maximal monotone operator and S be the semigroup generated by -A.

Proposition 4.2. For $x \in dom(A)$, $S(t)x \in dom(A)$, $t \geq 0$ and the mapping $t \to A^0S(t)x$ is continuous to the right on $[0,+\infty[$. The function $t \to S(t)x$ is differentiable to the right on $[0,+\infty[$ and we have

$$d^+/dt\, S(t)x + A^0S(t)x = 0, \quad t \geq 0.$$

This clarifies the anterior statement about the close relationship between monotone operators and nonlinear evolution equations.

Proof

We consider the equation

$$\frac{du_\lambda}{dt} + A_\lambda u_\lambda = 0, \quad u_\lambda(0) = x$$

which has a unique solution $u_\lambda \in C^1([0,\infty[;X)$ since A_λ (the Yosida approximation of A) is lipschitzian.

Let $h > 0$ be given, then

$$\left(\frac{du_\lambda}{dt}(t+h) - \frac{du_\lambda}{dt}(t), u_\lambda(t+h) - u_\lambda(t)\right)_X \leq 0, \quad \forall\, t \geq 0,$$

that is

$$\frac{1}{2}\frac{d}{dt}|u_\lambda(t+h) - u_\lambda(t)|_X^2 \leq 0.$$

We get that

$$|u_\lambda(t+h) - u_\lambda(t)|_X \leq |u_\lambda(h) - u_\lambda(0)|_X, \quad t \geq 0,$$

and, dividing by $h \to 0$, we obtain that $\left|\frac{du_\lambda}{dt}(t)\right|_X \leq \left|\frac{du_\lambda}{dt}(0)\right|_X$.

Using again the approximating equation, we infer

$$|A_\lambda u_\lambda(t)|_X = \left|\frac{du_\lambda}{dt}(t)\right|_X \leq \left|\frac{du_\lambda}{dt}(0)\right|_X = |A_\lambda x|_X \leq |A^o x|_X$$

by the properties of A_λ.

We subtract two equations corresponding to the parameters λ, μ and we multiply by $u_\lambda - u_\mu$. Using the identity

$$(A_\lambda u_\lambda - A_\mu u_\mu, u_\lambda - u_\mu)_X = (A_\lambda u_\lambda - A_\mu u_\mu, J_\lambda u_\lambda - J_\mu u_\mu)_X +$$
$$+ (A_\lambda u_\lambda - A_\mu u_\mu, \lambda A_\lambda u_\lambda - \mu A_\mu u_\mu)_X$$

we see that $\{u_\lambda\}$ is a Cauchy sequence in $C(0,T;X)$ for any $T < +\infty$ and we denote $u_\lambda(t) \to u(t)$ strongly in $C(0,T;X)$. By the above estimates, we have $u(t) \in \mathrm{dom}(A)$ for all $t \geq 0$ and $A_\lambda u_\lambda(t) \to A^o u(t)$ weakly in X.

As $\left\{\frac{du_\lambda}{dt}\right\}$ is bounded in $L^\infty(0,T;X)$, we may pass to the limit and see that u is the (unique) solution of the Cauchy problem $u(0) = x$ and

$$\frac{du}{dt} + Au = 0, \quad \text{a.e. } t \geq 0.$$

We show the continuity to the right of $A^o u(\cdot)$ and the differentiability to the right of $u(\cdot)$ for every $t \in [0,+\infty[$. It is enough to argue for $t = 0$. Let $t_n \to 0$, then $|A^o u(t_n)|_X \leq |A^o x|_X$, by the previously established estimates. We may assume that $A^o u(t_n) \to \xi$ weakly in X and since $u \in C(0,T;X)$, we obtain $\xi = A^o x$. A well known criterion for strong convergence in Hilbert spaces yields that $A^o u(t_n) \to A^o x$ strongly in X.

Again, by the same estimate, we have

$$|u(t_0 + h) - u(t_0)|_X \leq h|A^o u(t_0)|_X, \quad \forall\, t_0 \geq 0, \ \forall\, h \geq 0.$$

Then, if t_0 is such that u is differentiable in t_0, we infer that, in fact, we have

$$\frac{du}{dt} + A^o u(t_0) = 0 \quad \text{a.e. } [0,+\infty[.$$

Integrating this inequality, it yields

$$\frac{u(t) - u(0)}{t} = \frac{1}{t}\int_0^t A^o u(s)ds$$

and, we may use the right continuity of $A^o u(\cdot)$ to infer the right differentiability of $u(\cdot)$ for all $t \geq 0$.

Finally, we remark that the mapping $S(t)x = u(t)$ is the (nonlinear) semigroup generated by $-A$ and the proof is finished.

In the next chapters we discuss control problems governed by various types of evolution equations with monotone operators: parabolic and hyperbolic equations, delay differential systems, variational inequalities, free boundary problems. For the existence, uniqueness or regularity results which will be needed, we refer directly to the literature, in order to keep the auxiliary material at a minimum level. However, in Chapter IV,§1, we discuss some results related to the two-phase Stefan problem, which are relatively recent. Now, we indicate two general results for abstract, nonlinear evolution equations.

Theorem 4.3. Let $A \subset X \times X$ be a maximal monotone operator, X be a Hilbert space and $[y_0, f] \in \text{dom}(A) \times L^1(0,T;X)$, $\omega \in R$. There exists a unique weak solution $y \in C(0,T;X)$, $y(t) \in \text{dom}(A)$ a.e. $[0,T]$, of the equation:

$$dy/dt + Ay(t) + \omega y(t) \ni f(t),$$
$$y(0) = y_0.$$

Furthermore, we have the estimate

$$|y|_{C(0,T;X)} \leq C(1 + |f|_{L^1(0,T;X)}),$$

where $C = C(y_0, T; \omega)$.

Here, the statement that y is a weak solution has to be understood in the sense that there exist $\{f_n\} \subset W^{1,1}(0,T;X)$, $\{y_n\} \subset W^{1,\infty}(0,T;X)$, such that:

i) $dy_n/dt + Ay_n(t) + \omega y_n(t) \ni f_n(t)$ a.e. $]0,T[$,
ii) $y_n \to y$ in $C(0,T;X)$,
iii) $y(0) = y_0$, $f_n \to f$ in $L^1(0,T;X)$.

The result is a modification of Theorem 2.3, V.Barbu [12], p.135 and we give an outline of the argument.

Proof

Let $\{f_n\}$ be choosen as above. We consider the approximating equation

$$dy_n/dt + Ay_n + \omega y_n \ni f_n,$$
$$y_n(0) = y_o,$$

which has a unique solution $y_n \in W^{1,\infty}(0,T;X)$, $y_n(t) \in dom(A)$ a.e. $[0,T]$, from Barbu [12], p.135.

Subtract two equations corresponding to the parameters n,m and multiply by $y_n - y_m$:

$$1/2\, d/dt |y_n - y_m|_X^2 + \omega |y_n - y_m|_X^2 \leq (f_n - f_m, y_n - y_m)_X.$$

Therefore

$$|y_n(t) - y_m(t)|_X^2 \leq 2\int_o^t e^{2\omega(s-t)} |f_n(s) - f_m(s)|_X |y_n(s) - y_m(s)|_X ds.$$

By Proposition 2.9., we get

$$|y_n(t) - y_m(t)|_X \leq \int_o^t e^{2\omega(s-t)} |f_n(s) - f_m(s)|_X ds$$

and it yields $y_n \to y$ in $C(0,T;X)$.

For the final estimate, we multiply the approximating equation by $y_n - y_o$ and we proceed similarly to obtain

$$|y_n(t) - y_o|_X \leq \int_o^t e^{2\omega(s-t)} |f_n - \omega y_o - Ay_o|_X.$$

The uniqueness of the weak solution is a direct consequence of the monotonicity of A.

Consider now $V \subset H \subset V^*$, Hilbert spaces, with continuous, dense embedding, and a family of linear operators $A(t) : V \to V^*$, $t \in [0,T]$, which satisfy the conditions:

j) for any $y \in V$, the mapping $t \to A(t)y$ is strongly measurable on $[0,T]$;

jj) for any $t \in [0,T]$, $A(t)$ is continuous and there is $C>0$ such that

$$|A(t)|_{L(V,V^*)} \leq C \quad \text{a.e.}[0,T];$$

jjj) $(A(t)y,y)_{V \times V^*} + \alpha |y|_H^2 \geq \omega |y|_V^2$, $\omega > 0$, $y \in V$, a.e. $t \in [0,T]$.

Theorem 4.4. Let $f \in L^2(0,T;V^*)$ and $y_o \in H$; then there is a unique function $y \in C(0,T;H) \cap L^2(0,T;V)$, $dy/dt \in L^2(0,T;V^*)$ such that

$$dy/dt(t) + A(t)y(t) = f(t) \quad \text{a.e.}]0,T[,$$
$$y(0) = y_o.$$

If A is independent of t, $f \in L^2(0,T;H)$ and $Ay_0 \in H$, then it is possible to show that $y \in C(0,T;V)$, $dy/dt \in L^2(0,T;H)$, Barbu and Precupanu [14].

We close this section with a regularity result in $L^p(Q)$, $2 \leq p < +\infty$, for parabolic variational inequalities.

The concept of variational inequality is due to Lions and Stampacchia [73] and, roughly speaking, it corresponds to the case when the maximal monotone operator, which appears in the equation, is just a subdifferential.

Consider the parabolic variational inequality:

$$\text{(p)} \qquad \begin{aligned} & y_t - \Delta y + \beta(y) \ni f && \text{a.e. } Q = \Omega \times \,]0,T[, \\ & y(0,x) = y_0(x) && \text{a.e. } \Omega, \\ & y(t,x) = 0 && \text{a.e. } \Sigma = \Gamma \times [0,T], \end{aligned}$$

where $\beta \subset R \times R$ is a maximal monotone graph (see Example 2, § 3.1) and $f \in L^p(Q)$, $y_0 \in W_0^{1,p}(\Omega)$.

Theorem 4.5. Assume the above conditions hold together with the compatibility conditions $0 \in \mathrm{dom}\,\beta$, $y_0(x) \in \mathrm{dom}\,\beta$ a.e. Ω and there is $\gamma(x) \in \beta(y_0(x))$ a.e. Ω, such that $\gamma \in L^p(\Omega)$. Then (p) has a unique solution $y \in W^{2,1,p}(Q)$.

Proof

First we remark that, by modifying f with a constant, we may assume $0 \in \beta(0)$.

Let β_ε be the Yosida approximation of β. By the L^2 - theory of (p), Barbu [12], p.203, it is known that the approximating equation

$$\text{(p.a.)} \qquad \begin{aligned} & y_t - \Delta y + \beta_\varepsilon(y) = f && \text{a.e. } Q, \\ & y(0,x) = y_0(x) && \text{a.e. } \Omega, \\ & y(t,x) = 0 && \text{a.e. } \Sigma, \end{aligned}$$

has a unique solution $y_\varepsilon \in W^{2,1,2}(Q)$.

By the regularity results for linear parabolic equations, Solonnikov [103], and by Thm.1.3, an iterative argument based on the Lipschitz property of β_ε, shows that $y_\varepsilon \in W^{2,1,p}(Q)$.

Multiply (p.a.) by $\gamma_\varepsilon(y_\varepsilon) = |\beta_\varepsilon(y_\varepsilon)|^{p-2}\beta_\varepsilon(y_\varepsilon)$ and integrate over Q:

$$\int_0^T \int_\Omega \gamma_\varepsilon(y_\varepsilon)(y_\varepsilon)_t dxdt - \int_0^T \int_\Omega \Delta y_\varepsilon \gamma_\varepsilon(y_\varepsilon) dxdt + \int_Q |\beta_\varepsilon(y_\varepsilon)|^p dxdt =$$

$$= \int_Q f\, \gamma_\varepsilon(y_\varepsilon) dxdt \leq |f|_{L^p(Q)} |\beta_\varepsilon(y_\varepsilon)|^{p/p'}_{L^p(Q)},$$

where $1/p + 1/p' = 1$.

The mapping γ_ε is monotone and $\gamma_\varepsilon(0) = 0$ as $0 \in \beta(0)$. Then, the indefinite integral of γ_ε, denoted $\tilde{j}_\varepsilon(y) = \int_0^y \gamma_\varepsilon(s)ds$, attains its minimum value in 0 and we have $\tilde{j}_\varepsilon(y) \geq 0$ for all $y \in R$.

We consider the inequalities:

$$\int_0^T \int_\Omega \gamma_\varepsilon(y_\varepsilon)(y_\varepsilon)_t dxdt = \int_\Omega \tilde{j}_\varepsilon(y_\varepsilon(T,x))dx - \int_\Omega \tilde{j}_\varepsilon(y_0(x))dx \geq - \int_\Omega \tilde{j}_\varepsilon(y_0(x))dx,$$

$$-\int_0^T\int_\Omega \Delta y_\varepsilon \gamma_\varepsilon(y_\varepsilon)dxdt = \int_0^T\int_\Omega \text{grad } y_\varepsilon \text{ grad } \gamma_\varepsilon(y_\varepsilon)dxdt -$$

$$-\int_0^T\int_\Gamma \partial y_\varepsilon/\partial n \, \gamma_\varepsilon(y_\varepsilon)d\sigma dt = \int_0^T\int_\Omega \text{grad } y_\varepsilon \text{ grad}\gamma_\varepsilon(y_\varepsilon)dxdt \geq 0$$

and we obtain the estimate

$$-\int_\Omega \tilde{j}_\varepsilon(y_0(x))dx + \int_Q |\beta_\varepsilon(y_\varepsilon)|^p dxdt \leq |f|_{L^p(Q)} |\beta_\varepsilon(y_\varepsilon)|^{p/p'}_{L^p(Q)}.$$

By the compatibility hypothesis, we get

$$\int_\Omega \tilde{j}_\varepsilon(y_0(x))dx \leq \left|\int_\Omega\left\{\int_0^{y_0(x)} |\beta_\varepsilon(s)|^{p-2}\beta_\varepsilon(s)ds\right\}dx\right| \leq \int_\Omega |y_0(x)|$$

$$\cdot|\beta_\varepsilon(y_0(x))|^{p-1}dx \leq \int_\Omega |y_0(x)||\gamma(x)|^{p-1}dx \leq C.$$

Finally, we deduce that $\{\beta_\varepsilon(y_\varepsilon)\}$ is bounded in $L^p(Q)$ with respect to $\varepsilon > 0$. Using again the regularity results for the linear problem, it yields that $\{y_\varepsilon\}$ is bounded in $W^{2,1,p}(Q)$. One may easily pass to the limit in (p.a.) and finish the proof.

Remark 4.6. Let us consider the abstract evolution variational inequality

$$dy/dt + \partial\varphi(y) \ni f.$$

By the definition of the subdifferential, it may be written

$$(dy/dt, y-v)_X + \varphi(y) - \varphi(v) \leq (f, y-v)_{X'}, \qquad v \in X.$$

If φ is the indicator function of a convex, closed subset $C \subset X$, this is equivalent with $y \in C$ and

$$(dy/dt,\ y-v)_X \leq (f, y-v)_{X'}, \qquad v \in C.$$

In applications to partial differential equations, the set C may be indicated in various ways:

(α) conditions on the unknown function

$$C = \{y \in L^2(Q);\ y(t,x) \in \tilde{C} \quad \text{a.e.}Q\},$$

where $\tilde{C}$ is a convex, closed subset in R or R^N;

(β) conditions on the gradient

$$C = \{y \in L^2(0,T;H^1(\Omega));\ \text{grad } y(t,x) \in \tilde{C} \ \text{ a.e.}Q\};$$

(γ) conditions on y_t

$$C = \{y \in W^{1,2}(0,T;L^2(\Omega));\ y_t(t,x) \in \tilde{C} \ \text{ a.e. } Q\};$$

(δ) conditions on the trace of y on Σ

$$C = \{y \in L^2(0,T;H^1(\Omega));\ y(t,x) \in \tilde{C}\ \text{a.e.}\ \Sigma\},$$

etc. It is understood that, in each situation, X is a function space appropriately choosen.

We say that we have unilateral conditions (in the domain) on y, grad y, y_t, respectively unilateral conditions on the boundary. The variational inequality (p) is of type (α). Several examples are given in Chapter III, 1. In Chapter IV, § 1 we show the relationship between variational inequalities of the form (γ) and the two - phase Stefan problem.

A thorough study of the variational inequalities of different types may be found in the books by Kinderlehrer and Stampacchia [62], A.Friedman [47], H.Brezis [20], J.Naumann [81].

II. SEMILINEAR EQUATIONS

In this chapter we discuss optimal control problems governed by partial differential equations containing nonlinear terms which are locally Lipschitzian. These problems are neither convex, nor smooth. The general procedure to obtain the optimality conditions is the adapted penalization method, introduced by V.Barbu [13]. It consists in the approximation of the given problem by a family of smooth optimization problems and it allows the characterization of all the optimal pairs.

The results presented here play an important role in the next chapters, in the study of optimal control problems with state equation involving unbounded nonlinear operators: variational inequalities, free boundary problems.

1. An abstract control problem

We present, in an abstract setting, a general approximating process of nonlinear control problems.

Let X,Y,Z,W be Hilbert spaces with $Y \subset Z$ algebraically and topologically. The spaces X, Z, W may be assumed as identified with their duals. Let $S : Z \to X$, $F : W \to Y$ be linear, bounded operators and $M : Z \to Z$ be a maximal monotone, possibly multivalued, operator.

We consider the problem:

(1.1) $\quad \text{Minimize} \int_0^T L(Sy(t),u(t))dt$

subject to $u \in L^2(0,T;W)$ and $y \in C(0,T;Z)$, satisfying

(1.2) $\quad y'(t) + My(t) + \nu y(t) \ni Fu(t) \quad \text{in }]0,T[,$

(1.3) $\quad y(0) = y_0$

Here we assume that $L : X \times W \to]-\infty,+\infty]$ is a convex, lower semicontinuous, proper function with finite Hamiltonian (Ch.I, §3.2) and $y_0 \in \text{dom}(M)$, $\nu \in R$.

By Thm.4.3, Ch.I, it yields that for any $u \in L^2(0,T;W)$, equation (1.2), (1.3) has a unique weak solution $y \in C(0,T;Z)$. We denote by $\theta : L^2(0,T;Y) \to L^2(0,T;Z)$ the mapping from the right - - hand side to the solution, defined by (1.2), (1.3).

Let $M^\varepsilon : Z \to Z$, $\varepsilon > 0$, be a family of single valued, maximal monotone operators and $\theta_\varepsilon : L^2(0,T;Y) \to L^2(0,T;Z)$ be the mapping given by (1.2), (1.3) when M is replaced by M^ε.

The main hypotheses of this section are:

(a) $\theta_\varepsilon \circ F : L^2(0,T;W) \to C(0,T;Z)$ is completely continuous, uniformly with respect to ε from a neighbourhood of the origin; that is $u_n \to u$ weakly in $L^2(0,T;W)$ implies:

$$\theta_\varepsilon(Fu_n) \to \theta_\varepsilon(Fu) \text{ strongly in } C(0,T;Z),$$

uniformly with respect to ε in the given neighbourhood.

(b) $S \circ \theta_\varepsilon : L^2(0,T;Y) \to L^2(0,T;X)$ is Gateaux differentiable for any $\varepsilon > 0$.

(c) θ_ε approximates θ uniformly

$$|\theta_\varepsilon(f)(t) - \theta(f)(t)|_Z \leq \delta(\varepsilon), \quad t \in [0,T],$$

where $\delta(\varepsilon) \to 0$ when $\varepsilon \to 0$, uniformly with respect to f in bounded subsets of $L^2(0,T;Y)$.

Proposition 1.1. Suppose that L is coercive:

$$(1.4) \qquad \lim_{|u|_{L^2(0,T;W)} \to \infty} \int_0^T L(Sy, u) = +\infty$$

for y given by (1.2), (1.3). Then, the problem (1.1) - (1.3) has at least one optimal pair, denoted $[y^*, u^*]$.

Proof

Let $\{u_n\}$ be a minimizing sequence for (1.1). By (1.4) we see that $\{u_n\}$ is bounded in $L^2(0,T;W)$, so, on a subsequence, we have $u_n \to u$ weakly in $L^2(0,T;W)$. Then

$$\begin{aligned} |\theta(Fu_n) - \theta(Fu)|_{C(0,T;Z)} &\leq |\theta_\varepsilon(Fu_n) - \theta(Fu_n)|_{C(0,T;Z)} + \\ &+ |\theta_\varepsilon(Fu) - \theta(Fu)|_{C(0,T;Z)} + |\theta_\varepsilon(Fu_n) - \theta_\varepsilon(Fu)|_{C(0,T;Z)} \leq \\ &\leq 2\delta(\varepsilon) + |\theta_\varepsilon(Fu_n) - \theta_\varepsilon(Fu)|_{C(0,T;Z)}, \end{aligned}$$

since $\{Fu_n\}$ is bounded in $L^2(0,T;Y)$.

By hypotheses (a) and (c) it yields that $y_n = \theta(Fu_n) \to y = \theta(Fu)$ strongly in $C(0,T;Z)$ and the weak lower semicontinuity of L on X x W achieves the proof.

Remark 1.2. If L is a quadratic cost functional

$$L(Sy,u) = 1/2\,|Sy - y_d|_X^2 + 1/2\,|u|_W^2,$$

with $y_d \in L^2(0,T;X)$, then (1.4) is satisfied. If $U_{ad} \subset W$ is a closed, convex bounded subset and control constraints of the type $u \in U_{ad}$ are given, then we redefine L by $+\infty$ outside $X \times U_{ad}$ and again the coercivity condition (1.4) is fulfilled.

Let us consider the approximating problem

$$(1.5) \quad \text{Minimize}\left\{\int_0^T L^\varepsilon(Sy,u)dt + 1/2\int_0^T |u - u^*|_W^2 dt\right\}$$

subject to

(1.6) $y'(t) + M^{\varepsilon} y(t) + \nu y(t) = Fu(t)$ in $]0,T[$,
$y(0) = y_0$.

We denote $L^{\varepsilon} = L_{\delta(\varepsilon)}$, the regularization of the convex function L of order $\delta(\varepsilon)$, Ch.I, § 3.1. The functional (1.5) is coercive, even in the absence of the condition (1.4), since L^{ε} are majorized from below by an affine function, independent of ε, and θ_{ε} is sublinear according to Thm.4.3, Ch.I. Then Proposition 1.1 gives the existence of at least one optimal pair, $[y_{\varepsilon}, u_{\varepsilon}]$, for the problem (1.5),(1.6).

Lemma 1.3. For every $\varepsilon > 0$ there is $p_{\varepsilon} \in L^2(0,T;Y^*)$ such that

(1.7) $p_{\varepsilon} = -[\nabla(S \circ \theta_{\varepsilon})(Fu_{\varepsilon})]^* \partial_1 L^{\varepsilon}(Sy_{\varepsilon}, u_{\varepsilon})$,

(1.8) $F^* p_{\varepsilon} = \partial_2 L^{\varepsilon}(Sy_{\varepsilon}, u_{\varepsilon}) + u_{\varepsilon} - u^*$.

Here F^*, $[\ \ ]^*$ denote the adjoint operators.

Proof

Since L^{ε} is Frechet differentiable and $S \circ \theta_{\varepsilon}$ is Gateaux differentiable, we deduce that the functional

$$\Pi_{\varepsilon}(u) = \int_0^T L^{\varepsilon}(Sy,u)dt + 1/2\int_0^T |u - u^*|_W^2 dt,$$

where $y = \theta_{\varepsilon}(Fu)$, is Gateaux differentiable on $L^2(0,T;W)$, R.A.Tapia [110].

As u_{ε} is a minimum point for Π_{ε}, we get $\nabla\Pi_{\varepsilon}(u_{\varepsilon})v = 0$, $\forall v \in L^2(0,T;W)$. An elementary calculation, based on the chain rule, gives

$$\int_0^T (\partial_1 L^{\varepsilon}(Sy_{\varepsilon}, u_{\varepsilon}), \nabla(S \circ \theta_{\varepsilon})(Fu_{\varepsilon})Fv)_X dt +$$
$$+ \int_0^T (\partial_2 L^{\varepsilon}(Sy_{\varepsilon}, u_{\varepsilon}) + u_{\varepsilon} - u^*, v)_W dt = 0$$

for all $v \in L^2(0,T;W)$.

We define p_{ε} by (1.7), and (1.8) is an immediate consequence of the above equality.

Lemma 1.4. When $\varepsilon \to 0$, we have:

(1.9) $y_{\varepsilon} \to y^*$ strongly in $C(0,T;Z)$

(1.10) $u_{\varepsilon} \to u^*$ strongly in $L^2(0,T;W)$.

Proof

By the minimum property of u_{ε}, we obtain

$$\int_0^T L^{\varepsilon}(Sy_{\varepsilon}, u_{\varepsilon})dt + 1/2\int_0^T |u_{\varepsilon} - u^*|_W^2 dt \leq$$
$$\leq \int_0^T L^{\varepsilon}(S \circ \theta_{\varepsilon}(Fu^*), u^*)dt.$$

The definition of the Yosida regularization (§ 3.1., Ch.I) implies

$$L^{\varepsilon}(S\circ\theta_{\varepsilon}(Fu^*)(t),\ u^*(t)) \leq L(Sy^*(t),\ u^*(t)) +$$
$$+ C|\theta_{\varepsilon}(Fu^*)(t) - \theta(Fu^*)(t)|_Z^2/2\delta(\varepsilon), \quad \text{a.e.} t\in[0,T].$$

Assumption (c) yields

$$\limsup_{\varepsilon\to 0}\left\{\int_0^T L^{\varepsilon}(Sy_{\varepsilon}, u_{\varepsilon})dt + 1/2\int_0^T|u_{\varepsilon} - u^*|_W^2 dt \leq \int_0^T L(Sy^*, u^*)dt. \tag{1.11}$$

We have already remarked that the functionals Π_{ε} are coercive, uniformly with respect to $\varepsilon > 0$. Therefore, (1.11) shows that $\{u_{\varepsilon}\}$ is bounded in $L^2(0,T;W)$ and, on a convenient subsequence, we have $u_{\varepsilon}\to u_0$ weakly in $L^2(0,T;W)$.

By the inequality

$$|\theta_{\varepsilon}(Fu_{\varepsilon}) - \theta(Fu_0)|_{C(0,T;Z)} \leq |\theta_{\varepsilon}(Fu_{\varepsilon}) - \theta_{\varepsilon}(Fu_0)|_{C(0,T;Z)} +$$
$$+|\theta_{\varepsilon}(Fu_0) - \theta(Fu_0)|_{C(0,T;Z)} \leq \delta(\varepsilon) + |\theta_{\varepsilon}(Fu_{\varepsilon}) - \theta_{\varepsilon}(Fu_0)|_{C(0,T;Z)}$$

and by hypotheses (a), (c), we get that $y_{\varepsilon} = \theta_{\varepsilon}(Fu_{\varepsilon})\to\tilde{y} = \theta(Fu_0)$ strongly in $C(0,T;Z)$.

By the definition of L^{ε} and (1.11), we see that $J_{\delta(\varepsilon)}(Sy_{\varepsilon}, u_{\varepsilon}) = (I + \delta(\varepsilon)\partial L)^{-1}(Sy_{\varepsilon}, u_{\varepsilon}) \to [S\tilde{y}, u_0]$ weakly in $L^2(0,T;X) \times L^2(0,T;W)$.

When $\varepsilon\to 0$ in (1.11), by the weak lower semicontinuity of the cost functional and by Thm 3.6, Ch. I, we obtain

$$\int_0^T L(S\tilde{y}, u_0)dt + 1/2\int_0^T|u_0 - u^*|_W^2 dt \leq \int_0^T L(Sy^*, u^*)dt. \tag{1.12}$$

But $[y^*, u^*]$ is an optimal pair for the problem (1.1) - (1.3), therefore $u_0 = u^*$, $\tilde{y} = y^*$ and $u_{\varepsilon}\to u^*$ strongly in $L^2(0,T;W)$.

Lemma 1.5. The Gateaux differential satisfies

$$|[\nabla(S\circ\theta_{\varepsilon})(w)v](t)|_X \leq C\int_0^t|v(s)|_Y ds, \quad w,v\in L^2(0,T;Y),\ t\in[0,T], \tag{1.13}$$

$$|[\nabla(S\circ\theta_{\varepsilon})(w)^*v](t)|_{Y^*} \leq C\int_t^T|v(s)|_X ds, \quad t\in[0,T],\ w\in L^2(0,T;Y),\ v\in L^2(0,T;X). \tag{1.14}$$

where C is a constant independent of $\varepsilon > 0$.

Proof

We have

$$\nabla(S\circ\theta_{\varepsilon})(w)v = \lim_{\lambda\to 0}(Sy_{\lambda} - Sy)/\lambda$$

in $L^2(0,T;X)$, where

$$y_{\lambda}'(t) + M^{\varepsilon}y_{\lambda}(t) + \nu y_{\lambda}(t) = w(t) + \lambda v(t) \quad \text{in }]0,T[$$
$$y'(t) + M^{\varepsilon}y(t) + \nu y(t) = w(t) \quad \text{in }]0,T[$$
$$y_{\lambda}(0) = y(0) = y_0.$$

Subtract the two equations, multiply by y - y and integrate over [0,s]:

(1.15) $$1/2|y_\lambda(s) - y(s)|_Z^2 + \nu\int_0^s |y_\lambda(\tau) - y(\tau)|_Z^2 d\tau \leq \lambda\int_0^s (v(\tau), y_\lambda(\tau) - y(\tau))_Z d\tau.$$

We multiply by $e^{2\nu s}$ and integrate over [0,t]. After a short calculation, we get

(1.16) $$\int_0^t |y_\lambda(s) - y(s)|_Z^2 ds \leq \lambda/\nu \int_0^t (v(s), y_\lambda(s) - y(s))_Z ds - \lambda/\nu \int_0^t e^{2\nu(s-t)}(v(s), y_\lambda(s) - y(s))_Z ds.$$

If $\nu \geq 0$ we can use directly (1.15) and the Gronwall inequality to obtain (1.13). Assume that $\nu < 0$ and multiply by ν in (1.16). Taking into account (1.15) again, we have

$$|y_\lambda(s) - y(s)|_Z^2 \leq C\lambda\int_0^t |v(s)|_Y \cdot |y_\lambda(s) - y(s)|_Z ds.$$

The <u>Proposition 1.9</u> gives (1.13). In particular, it yields that the operator $\nabla(S\circ\theta_\varepsilon)(w) : L^2(0,T;Y) \to L^2(0,T;X)$ may be extended by continuity to the whole space $L^1(0,T;Y)$. The adjoint operator $\nabla(S\circ\theta_\varepsilon)(w)^* : L^2(0,T;X) \to L^\infty(0,T;Y^*)$ satisfies:

$$|\nabla(S\circ\theta_\varepsilon)(w)^* v|_{L^\infty(t,T;Y^*)} = \sup_{|p|_{L^1(t,T;Y)} \leq 1} \int_t^T (\nabla(S\circ\theta_\varepsilon)(w)^* v, p)_{Y\times Y^*} ds =$$

$$= \sup_{|p|_{L^1(t,T;Y)} \leq 1} \int_0^T (v, (S\circ\theta_\varepsilon)(w)\tilde{p})_X ds \leq$$

$$\leq \sup_{|p|_{L^1(t,T;Y)} \leq 1} \int_0^T |v(s)|_X |\nabla(S\circ\theta_\varepsilon)(w)\tilde{p}|_X ds \leq C\int_t^T |v(s)|_X ds.$$

Above $v \in L^2(0,T;X)$, $\tilde{p}$ is the extension of p to [0,T] by 0 and we use (1.13).

<u>Theorem 1.6. There exist</u> p^* $L\ (0,T;Y^*)$, q^* $L^2(0,T;X)$ <u>such that, for</u> $\varepsilon \to 0$ <u>on a subsequence, we have:</u>

(1.17) $p_\varepsilon \to p^*$ <u>weakly* in</u> $L^\infty(0,T;Y^*)$,

(1.18) $\partial_1 L^\varepsilon(Sy_\varepsilon, u_\varepsilon) \to q^*$ <u>weakly in</u> $L^1(0,T;X)$,

(1.19) $[q^*(t), F^* p^*(t)] \in \partial L(Sy^*(t), u^*(t))$ a.e. [0,T].

<u>Proof</u>

By the definition of the subdifferential, we may write

$$(\partial_1 L^\varepsilon(Sy_\varepsilon, u_\varepsilon), Sy_\varepsilon - Sy^* - \rho w)_X + (\partial_2 L^\varepsilon(Sy_\varepsilon, u_\varepsilon), u_\varepsilon - v_0)_W \geq$$
$$\geq L^\varepsilon(Sy_\varepsilon, u_\varepsilon) - L^\varepsilon(Sy^* + \rho w, v_0),$$

for all $w \in X$, $\rho > 0$. We take $|w|_X = 1$ and ε sufficiently small and obtain via (1.8), (1.9):

$$\rho/2\,|\partial_1 L^\varepsilon(Sy_\varepsilon(t),u_\varepsilon(t))|_X \le$$
$$\le (F^* p_\varepsilon + u^* - u_\varepsilon, u_\varepsilon - v_0)_W + \sup_{|w|_X = 1} L^\varepsilon(Sy^* + \rho w, v_0) - L^\varepsilon(Sy_\varepsilon, u_\varepsilon).$$

By the finite Hamiltonian hypothesis, the Hamiltonian $\mathcal{H}(y,p)$ of L and its subdifferential are locally bounded in X x W. Let $v_0(t)$ be a measurable section of $\partial_p\mathcal{H}(Sy^*(t) + \rho w, 0)$. For the existence of such a v_0 we quote V.Barbu, Th.Precupanu [14], Ch. IV.

If $|w|_X = 1$ and ρ is sufficiently small, we have $|v_0(t)|_W \le$ ct. and $L(Sy^*(t) + \rho w, v_0(t)) = -\mathcal{H}(Sy^*(t) + \rho w, 0) \le$ ct. a.e. [0,T]. Here we use the fact that L is the partial conjugate of $\mathcal{H}$ and the definition of v_0. Then, we get the estimate

$$\rho/2\,|\partial_1 L^\varepsilon(Sy_\varepsilon(t),u_\varepsilon(t))|_X \le (|F^* p_\varepsilon(t)|_W + |u_\varepsilon(t) - u^*(t)|_W)(C + |u_\varepsilon(t)|_W) + C. \tag{1.20}$$

From (1.14) and (1.7) it yields

$$|p_\varepsilon(t)|_{Y^*} \le C \int_t^T |\partial_1 L^\varepsilon(Sy_\varepsilon(s),u_\varepsilon(s))|_X ds.$$

Combining this with (1.20), the Gronwall lemma gives $|p_\varepsilon(t)|_{Y^*} \le C$ a.e.[0,T]. We take into account this in (1.20) and we obtain

$$|\partial_1 L^\varepsilon(Sy_\varepsilon(t),u_\varepsilon(t))|_X \le C(1 + |u_\varepsilon(t) - u^*(t)|_W)(1 + |u_\varepsilon(t)|_W) \quad \text{a.e. } [0,T]. \tag{1.21}$$

The Dunford - Pettis criterion (§1, Ch.I) implies that, on a subsequence $\partial_1 L^\varepsilon(Sy_\varepsilon, u_\varepsilon) \to q^*$ weakly in $L^1(0,T;X)$. Obviously, we may take $p_\varepsilon \to p^*$ weakly* in $L^\infty(0,T;Y^*)$.

Let σ be fixed in $L^\infty(0,T)$. By (1.21), we have

$$\int_0^T |\partial_1 L^\varepsilon(Sy_\varepsilon, u_\varepsilon)|_X \sigma dt \le \int_0^T C(1 + |u_\varepsilon(t) - u^*(t)|_W)(1 + |u_\varepsilon(t)|_W)|\sigma| dt.$$

When $\varepsilon \to 0$, we deduce

$$\int_0^T |q^*|_X \sigma dt \le C \int_0^T (1 + |u^*(t)|_W)|\sigma| dt \le C\,|1 + |u^*(t)|_W|_{L^2(0,T)} |\sigma|_{L^2(0,T)}.$$

It follows that $q^* \in L^2(0,T;X)$.

The inclusion (1.19) is a standard consequence of the demiclosedness of the maximal monotone operators and of (1.17), (1.18), (1.8), (1.9), (1.10).

Remark 1.7. The results of this section remain valid if u^* is only a local minimum for the functional (1.1). The approximate problem (1.5) has to be replaced by

$$\text{Minimize}\left\{ \int_0^T L^\varepsilon(Sy(t),u(t))dt + \eta \int_0^T |u(t) - u^*(t)|_W^2 dt \right., \tag{1.5)'$$

where η is a positive constant.

As it may be seen from the proof, the term $\eta \int_0^T |u_\varepsilon - u^*|_W^2$ is bounded, uniformly with respect to ε and η. If η is sufficiently large, it yields that $\{u_\varepsilon\}$ is in the neighbourhood of u^*, where the minimum property holds. The reasoning follows the same steps as above.

Remark 1.8. The relation (1.19) corresponds to the so called "maximum principle" for the optimal control problem (1.1) - (1.3).

Remark 1.9. We use Thm 1.6 in many situations, both in this chapter and in the next ones. For this, we write the optimization problems which we investigate, in the form (1.1) -(1.3). That is, we define the spaces X,Y,Z,W and the operators F, M ,S, in each case. Next, we choose a convenient approximation M^ε, which generally doesn't coincide with the Yosida approximation of the operator M. A difficult task is to check the conditions (a), (b), (c). Sometimes for hyperbolic or parabolic problems, we shall need nonstandard existence and regularity results.

In the last step we pass to the limit in (1.7), in order to obtain the adjoint equation. This is not supplied by Thm 1.6 since, generally, it involves to pass to the limit in the product of two weakly convergent sequences and has to be studied separately, in each application.

2. Parabolic problems

Let us consider $X = Y = Z$ and $M = \partial\varphi$, where $\varphi : Z\ ]-\ ,+\ ]$ is a convex, lower semicontinuous, proper function. The operator $S : Z \to Z$ is the identity. This choice corresponds to parabolic problem and was studied by V.Barbu, [11], Ch.5. Our treatment starts from the abstract scheme developped in §1. To fix the ideas, we examine the following control problem:

(2.1) $\quad$ Minimize $\int_0^T L(y,u)dt$

over all $u \in L^2(0,T;W)$ and y given by

(2.2) $\quad y_t - \Delta y + \beta(y) = Fu \qquad$ a.e. $Q =]0,T[\times \Omega$,

(2.3) $\quad y(0,x) = y_0(x) \qquad$ a.e. Ω,

(2.4) $\quad y(t,x) = 0 \qquad$ a.e. $\Sigma = \Gamma \times [0,T]$.

Here Ω is a bounded regular domain in R^N, $y_0 \in H_0^1(\Omega)$ and $\beta : R \to R$ is a locally Lipschitzian monotone mapping, satisfying the growth condition:

(2.5) $\quad |\dot\beta(y)| \leq C(|\beta(y)| + |y| + 1) \qquad$ a.e. R.

For convenience, we also ask $0 = \beta(0)$.

The condition (2.5) will be of frequent use in this chapter and it admits a large class of examples, including polynomials or exponentials.

We take $Z = L^2(\Omega)$ and $\varphi : Z \to]-\infty,+\infty]$ given by

$$(2.6) \qquad \varphi(y) = \begin{cases} 1/2\int_\Omega |\mathrm{grad}\, y|^2 dx + \int_\Omega j(y)dx, & \text{if } y \in H_0^1(\Omega), j(y) \in L^1(\Omega), \\ +\infty, & \text{otherwise,} \end{cases}$$

where $j : R \to]-\infty,+\infty]$ is a convex, lower semicontinuous, proper functional such that $\beta = \partial j$ (Example 2, 3.1,Ch.I). Here $j(r) = \int_0^r \beta(s)ds$.

We have

$$\partial\varphi(y) = \{-\Delta y + \beta(y)\} \cap L^2(\Omega),$$
$$\mathrm{dom}(\partial\varphi) = H_0^1(\Omega) \cap H^2(\Omega).$$

Therefore, the problem (2.1) - (2.4) has the form (1.1) - (1.3) with $M = \partial\varphi$, φ from (2.6).

We remark that for any control $u \in L^2(0,T;W)$, the state equation (2.2) - (2.4) has a unique solution $y \in C(0,T;L^2(\Omega))$, $y_t \in L^2(Q)$, $\Delta y \in L^2(Q)$ and we have the estimate

$$(2.7) \qquad |y_t|_{L^2(Q)} + |\Delta y|_{L^2(Q)} + |y|_{C(0,T;L^2(\Omega))} \leq C(|Fu|_{L^2(Q)}, |y_0|_{H_0^1(\Omega)})$$

with $C(.\,,.)$ a mapping on $R \times R$, bounded on bounded sets (see Brezis [20], Ch.II).

To introduce the operators M^ε, $\varepsilon > 0$, we use a penalization - regularization technique, which appeared first, in this type of problems, in the work of J.P.Yvon [144]. We define:

$$(2.8) \qquad \beta^\varepsilon(y) = \int_R \beta_\varepsilon(y - \varepsilon\tau)\rho(\tau)d\tau, \quad y \in R,$$
$$(2.9) \qquad j^\varepsilon(y) = \int_R j_\varepsilon(y - \varepsilon\tau)\rho(\tau)d\tau, \quad y \in R,$$

where β_ε is the Yosida approximation of β, j_ε is the regularization of the convex function j, and $\rho \in C_0^\infty(R)$ is a Friedrichs mollifier, that is $\rho \geq 0$, $\mathrm{supp}\rho \subset [-1,1]$, $\rho(-\tau) = \rho(\tau)$, $\int_{-\infty}^{\infty} \rho(\tau)d\tau = 1$. Then $\beta^\varepsilon = \partial j^\varepsilon$ and we put $M^\varepsilon = \partial\varphi^\varepsilon$ with φ^ε given by (2.6) when j is replaced by j^ε.

Obviously β^ε is maximal monotone and Lipschitzian of constant $1/\varepsilon$. Moreover, by (2.5) and (2.8), we obtain:

$$(2.5)' \qquad |\dot\beta^\varepsilon(y)| \leq C(|\beta^\varepsilon(y)| + |y| + 1), \qquad y \in R,$$

with C a constant independent of ε. The argument uses essentially the monotonicity of β:

$$0 \leq \dot\beta_\varepsilon(y) = \dot\beta((I + \varepsilon\beta)^{-1}(y))/(1 + \varepsilon\dot\beta((I + \varepsilon\beta)^{-1}(y))) \leq$$
$$\leq C(|\beta((I + \varepsilon\beta)^{-1}(y))| + |(I + \varepsilon\beta)^{-1}(y)| + 1) \leq C(|\beta_\varepsilon(y)| + |y| + 1).$$

By (2.8), it yields

$$0 \leq \dot\beta^\varepsilon(y) \leq C\int_{-1}^{1}(|\beta_\varepsilon(y - \varepsilon\tau)| + |y - \varepsilon\tau| + 1)\rho(\tau)d\tau.$$

If $|y| \leq \varepsilon$, as β_ε is Lipschitzian of constant $1/\varepsilon$, we see that $|\dot\beta^\varepsilon(y)| \leq C(|y| + 1)$ and (2.5)' is proved. If $y < -\varepsilon$, by the monotonicity of β_ε and from $\beta_\varepsilon(0) = 0$, the above inequality becomes

$$0 \le \dot\beta^{\varepsilon}(y) \le C\int_{-1}^{1}(-\beta_{\varepsilon}(y-\varepsilon\tau)+|y|+1)\rho(\tau)d\tau =$$
$$= C(-\beta^{\varepsilon}(y)+|y|+1) \le C(|\beta^{\varepsilon}(y)|+|y|+1).$$

Here C are different constants independent of ε. The situation $y > \varepsilon$ is similar.

Now, the mapping $\theta_{\varepsilon}: L^2(Q) \to C(0,T;L^2(\Omega))$ is defined by (2.2)-(2.4) where β is replaced by β^{ε}. Since the estimate (2.7) is based only on the monotonicity of β, it is preserved for θ_{ε}, with $C(.,.)$ independent of ε. Then, one may infer by an argument similar to Th. 4.5, Ch. I that $\theta_{\varepsilon}: L^2(Q) \to C(0,T;L^2(\Omega))$ is completely continuous, uniformly with respect to ε. This yields condition (a) of § 1.

Let us now check the assumptions (b) and (c):

Lemma 2.1. *The operator* $\theta_{\varepsilon}: L^2(Q) \to L^2(Q)$ *is Gateaux differentiable and* $r = \nabla\theta_{\varepsilon}(f)g$, *with* $f,g \in L^2(Q)$, *satisfies*

$$r_t - \Delta r + \dot\beta^{\varepsilon}(\theta_{\varepsilon}(f))r = g \qquad \text{a.e. in } Q,$$
$$r(t,x) = 0 \qquad \text{a.e. in } \Sigma,$$
$$r(0,x) = 0 \qquad \text{a.e. in } \Omega.$$

Moreover $r \in L^2(0,T;H^2(\Omega)) \cap W^{1,2}(0,T;L^2(\Omega))$.

Proof

We denote $y^{\lambda} = \theta_{\varepsilon}(f+\lambda g)$, $y = \theta_{\varepsilon}(f)$, $z = (y^{\lambda} - y)/\lambda$ and we have

(2.10) $$y_t^{\lambda} - \Delta y^{\lambda} + \beta^{\varepsilon}(y^{\lambda}) = f + \lambda g \qquad \text{a.e. in } Q,$$

(2.11) $$y_t - \Delta y + \beta^{\varepsilon}(y) = f \qquad \text{a.e. in } Q,$$
$$y^{\lambda}(t,x) = y(t,x) = 0 \qquad \text{a.e. in } \Sigma,$$
$$y^{\lambda}(0,x) = y(0,x) = y_0(x) \qquad \text{a.e. in } \Omega.$$

We subtract (2.10) and (2.11) and divide by λ

(2.12) $$z_t^{\lambda} - \Delta z^{\lambda} + (\beta^{\varepsilon}(y^{\lambda}) - \beta^{\varepsilon}(y))/\lambda = g \qquad \text{a.e. } Q,$$
$$z^{\lambda}(t,x) = 0 \text{ a.e. } \Sigma,\ z^{\lambda}(0,x) = 0 \qquad \text{a.e. } \Omega.$$

Multiply (2.12) by z^{λ} and use the monotonicity of β^{ε} to obtain $\{z^{\lambda}\}$ bounded in $L^{\infty}(0,T;L^2(\Omega))$. As β^{ε} is Lipschitzian (ε is fixed), we get $\{\lambda^{-1}(\beta^{\varepsilon}(y^{\lambda}) - \beta^{\varepsilon}(y))\}$ bounded in $L^2(Q)$.

Again by (2.12), we have $\{z^{\lambda}\}$ bounded in $L^2(0,T;H^2(\Omega)) \cap W^{1,2}(0,T;L^2(\Omega))$. We denote r the strong limit of z^{λ} in $C(0,T;L^2(\Omega))$, $r = \nabla\theta^{\varepsilon}(f)g$. Obviously

$$(\beta^{\varepsilon}(y^{\lambda}) - \beta^{\varepsilon}(y))/\lambda \to \dot\beta^{\varepsilon}(y)r$$

a.e. Q, because $y^{\lambda} \to y$ strongly in $L^2(Q)$. But the convergence is, in fact, in the weak topology of $L^2(Q)$, due to the boundedness of $\{\lambda^{-1}(\beta^{\varepsilon}(y^{\lambda}) - \beta^{\varepsilon}(y))\}$. This is a consequence of the Egorov theorem, Ch. I, § 1. From the a.e. convergence, we obtain that $\forall\ \mu > 0\ \exists\ Q_{\mu} \subset Q$ such that mes $(Q - Q_\mu) < \mu$ and $\lambda^{-1}(\beta^\varepsilon(y^\lambda) - \beta^\varepsilon(y)) \to \dot\beta^\varepsilon(y)r$ uniformly in Q_μ. As $\lambda^{-1}(\beta^\varepsilon(y^\lambda) - \beta^\varepsilon(y))$ is weakly convergent in $L^2(Q)$, it is also weakly convergent in $L^2(Q_\mu)$ and we see that the weak limit is $\dot\beta^\varepsilon(y)r$. One may pass to the limit in (2.12) and finish the proof.

Lemma 2.2. We have the estimate

(2.13) $$|\theta_\varepsilon(f) - \theta(f)|^2_{L^\infty(0,T;L^2(\Omega))} + |\theta_\varepsilon(f) - \theta(f)|^2_{L^2(0,T;H^1_0(\Omega))} \le \varepsilon C(|f|_{L^2(Q)}, |y_0|_{H^1_0(\Omega)}),$$

where $f \in L^2(Q)$, and $C(.,.)$ is a real function, bounded on bounded subsets in $R \times R$.

Proof

Let $y^\varepsilon = \theta_\varepsilon(f)$, $y = \theta(f)$. By (2.7), which is also valid for θ_ε, uniformly with respect to ε, we conclude that $\{y^\varepsilon\}$ is bounded in $L^2(0,T;H^2(\Omega)) \cap W^{1,2}(0,T;L^2(\Omega))$ and $\{\beta^\varepsilon(y^\varepsilon)\}$ is bounded in $L^2(Q)$ by $C(|f|_{L^2(Q)}, |y_0|_{H^1_0(\Omega)})$ which appears in (2.7).

We subtract the equations corresponding to θ_ε, θ_δ, with ε, δ positive, and we multiply by $y^\varepsilon - y^\delta$:

$$1/2|y^\varepsilon(t) - y^\delta(t)|^2_{L^2(\ )} + \int_0^t |\mathrm{grad}\, y^\varepsilon - \mathrm{grad}\, y^\delta|^2_{L^2(\Omega)} dt +$$
$$+ \int_0^t \int_\Omega (\beta^\varepsilon(y^\varepsilon) - \beta^\delta(y^\delta))(y^\varepsilon - y^\delta) dx dt = 0.$$

According to the properties of the Yosida approximation, the last term may be rewritten as follows:

$$(\beta^\varepsilon(y^\varepsilon) - \beta^\delta(y^\delta))(y^\varepsilon - y^\delta) = \int_R (\beta_\varepsilon(y^\varepsilon - \varepsilon\tau) - \beta_\delta(y^\delta - \delta\tau))(y^\varepsilon - y^\delta)\rho(\tau)d\tau =$$
$$= \int_R (\beta_\varepsilon(y^\varepsilon - \varepsilon\tau) - \beta_\delta(y^\delta - \delta\tau))(\varepsilon\beta_\varepsilon(y^\varepsilon - \varepsilon\tau) -$$
$$- \delta\beta_\delta(y^\delta - \delta\tau) + J_\varepsilon(y^\varepsilon - \varepsilon\tau) - J_\delta(y^\delta - \delta\tau) + \varepsilon\tau - \delta\tau)\rho(\tau)d\tau \ge$$
$$\ge \int_R (\beta_\varepsilon(y^\varepsilon - \varepsilon\tau) - \beta_\delta(y^\delta - \delta\tau))(\varepsilon\beta_\varepsilon(y^\varepsilon - \varepsilon\tau) - \delta\beta_\delta(y^\delta - \delta\tau) + \varepsilon\tau - \delta\tau)\rho(\tau)d\tau,$$

where $J_\varepsilon(y) = (I + \varepsilon\beta)^{-1}(y)$.

We remark the inequalities:

(2.14) $$|\beta_\varepsilon(y^\varepsilon - \varepsilon\tau)| \le |\beta_\varepsilon(y^\varepsilon)| + 1 \le |\beta^\varepsilon(y^\varepsilon)| + 2, \quad > 0$$

since β is Lipschitzian of constant $1/\varepsilon$.

By the above computation, it yields

(2.15) $$1/2|y^\varepsilon(t) - y^\delta(t)|^2_{L^2(\Omega)} + \int_0^t |\mathrm{grad}(y^\varepsilon - y^\delta)|^2_{L^2(\Omega)} ds \le C(|f|_{L^2(Q)}, |y_0|_{H^1_0(\Omega)})(\varepsilon + \delta)$$

as $\{\beta^\varepsilon(y^\varepsilon)\}$ is bounded in $L^2(Q)$.

Let $\tilde y$ be the strong limit of y^ε in $C(0,T;L^2(\Omega))$. It is a standard argument to see that $\tilde y = \theta(f)$ and, passing to the limit $\delta \to 0$ in (2.15), to obtain (2.13) and to end the proof.

If the coercivity hypothesis (1.4) is imposed, then, by Proposition 1.1, we get the existence of at least one optimal pair $[y^*, u^*]$ for the problem (2.1)-(2.4).

The approximating optimal control problem reads as follows:

(2.16) $$\text{Minimize}\left\{\int_0^T L^\varepsilon(y,u)dt + 1/2\int_0^T |u-u^*|_W^2 dt\right\}$$

subject to

(2.17) $$y_t - \Delta y + \beta^\varepsilon(y) = Fu \qquad \text{a.e. } Q$$

and (2.3), (2.4), where L^ε is as in §1 with $\delta(\varepsilon) = C\,\varepsilon^{1/2}$.

We denote by $[y^\varepsilon, u_\varepsilon]$ an optimal pair for the problem (2.16), (2.17). By Thm.1.6, we get

Proposition 2.3. There is $p^\varepsilon \in C(0,T;L^2(\Omega))$ such that it satisfies the approximate optimality system

$$
\begin{aligned}
&(2.18)\quad p_t^\varepsilon + \Delta p^\varepsilon - \dot\beta^\varepsilon(y^\varepsilon)p^\varepsilon = \partial_1 L^\varepsilon(y^\varepsilon, u_\varepsilon) && \text{a.e. } Q,\\
&(2.19)\quad p^\varepsilon(t,x) = 0 && \text{a.e. } \Sigma,\\
&(2.20)\quad p^\varepsilon(T,x) = 0 && \text{a.e. } \Omega,\\
&\qquad\quad F^* p^\varepsilon = \partial_2 L^\varepsilon(y^\varepsilon, u_\varepsilon) + u_\varepsilon - u^* && \text{a.e. } [0,T],
\end{aligned}
$$

together with (2.17), (2.3), (2.4). Moreover $u_\varepsilon \to u^*$ strongly in $L^2(0,T;W)$, $y^\varepsilon \to y^*$ strongly in $C(0,T;L^2(\Omega))$, $\partial_1 L^\varepsilon(y^\varepsilon, u_\varepsilon) \to q^*$ weakly in $L^1(0,T;L^2(\Omega))$, $p^\varepsilon \to p^*$ weakly in $L^2(0,T;H_0^1(\Omega))$ and $[q^*, F^*p^*] \in \partial L(y^*, u^*)$ a.e. $[0,T]$.

Proof

We have to show that (1.7) may be expressed in the form (2.18)-(2.20), in this case. We remark that (2.18)-(2.20) has a unique solution $p^\varepsilon \in C(0,T;L^2(\Omega)) \cap L^2(0,T;H_0^1(\Omega))$, with $p_t^\varepsilon \in L^2(0,T;H^{-1}(\Omega))$, by the results for linear parabolic problems, Barbu and Precupanu [14], p.64. Since β^ε is Lipschitz, we deduce that $p^\varepsilon \cdot \dot\beta^\varepsilon(y^\varepsilon) \in L^2(Q)$, therefore the regularity of p^ε is maximal, in fact.

The desired conclusion is obtained multiplying (2.18) by r given in Lemma 2.1 and integrating by parts several times:

$$\int_Q \partial_1 L^\varepsilon(y^\varepsilon, u_\varepsilon) r\,dxdt = \int_Q r(p_t^\varepsilon + \Delta p^\varepsilon - \dot\beta^\varepsilon(y^\varepsilon)p^\varepsilon)dxdt =$$

$$= -\int_Q p^\varepsilon(r_t - \Delta r + \dot\beta^\varepsilon(y^\varepsilon)r)dxdt = -\int_Q p^\varepsilon g\,dxdt\,.$$

We know that $\{\partial_1 L^\varepsilon(y^\varepsilon, u_\varepsilon)\}$ is bounded in $L^1(0,T;L^2(\Omega))$ and that $\dot\beta^\varepsilon(y^\varepsilon) \geq 0$. Multiplying (2.18) by p^ε we see that $\{p^\varepsilon\}$ is bounded in $L^\infty(0,T;L^2(\Omega)) \cap L^2(0,T;H_0^1(\Omega))$ and the proof is finished.

The next theorem shows how to pass to the limit in (2.18)-(2.20) and completes the results of §1 in this respect.

Theorem 2.4. Under condition (2.5), $p^* \in L^2(0,T;H_0^1(\Omega))$ satisfies the adjoint system:

$$p_t^* + \Delta p^* - D\beta(y^*)p^* \ni q^* \quad \text{in } Q,$$
$$p^*(t,x) = 0 \quad \text{in } \Sigma,$$
$$p^*(T,x) = 0 \quad \text{in } \Omega,$$

and

$$[q^*(t), F^*p^*(t)] \in \partial L(y^*(t), u^*(t)) \quad \text{a.e. } [0,T].$$

Proof

By (2.7), we have $\{\beta^\varepsilon(y^\varepsilon)\}$ bounded in $L^2(Q)$ and condition (2.5) implies that $\{\dot\beta^\varepsilon(y^\varepsilon)\}$ is bounded in $L^2(Q)$. On a subsequence, again denoted ε, we obtain:

$$\dot\beta^\varepsilon(y^\varepsilon) \to \gamma^* \in D\beta(y^*),$$

by Thm.3.14, Ch.I and by $y^\varepsilon \to y^*$ strongly in $C(0,T;L^2(\Omega))$.

To determine the limit of the product $\dot\beta^\varepsilon(y^\varepsilon).p^\varepsilon$, we prove that $p^\varepsilon \to p^*$ strongly in $L^2(Q)$.

Because $\{p^\varepsilon\}$ is bounded in $L^\infty(0,T;L^2(\Omega)) \cap L^2(0,T;H_0^1(\Omega))$ and $\{\dot\beta^\varepsilon(y^\varepsilon)\}$ is bounded in $L^2(Q)$, we get $\{\dot\beta^\varepsilon(y^\varepsilon)p^\varepsilon\}$ bounded in $L^1(Q)$.

Take $s > N/2$. The Sobolev embedding theorem gives $H^s(\Omega) \subset C(\bar\Omega)$, therefore $L^1(\Omega) \subset H^s(\Omega)^*$ and (2.18) implies that $\{p_t^\varepsilon\}$ is bounded in $L^1(0,T;V^*)$, where $V = H^s(\Omega) \cap H_0^1(\Omega)$. Obviously $L^2(\Omega) \subset V^*$ compactly and the Helly-Foias compactness theorem (§1,Ch.I) yields that, on a subsequence

$$p^\varepsilon(t) \to p^*(t) \text{ strongly in } V^*, \quad t \in [0,T],$$

and $p^* \in BV(0,T;V^*)$. But $\{p^\varepsilon\}$ is bounded in $L^\infty(0,T;V^*)$ and, consequently, $p^\varepsilon \to p^*$ strongly in $L^2(0,T;V^*)$ by the Lebesgue's dominated convergence theorem.

Now, we use the Lions's lemma (Ch.I,§1) since $H_0^1(\Omega) \subset L^2(\Omega)$ compactly: for any $\lambda > 0$, there is $C(\lambda) > 0$ such that

$$|p^\varepsilon(t) - p^*(t)|_{L^2(\Omega)} \le \lambda |p^\varepsilon(t) - p^*(t)|_{H_0^1(\Omega)} + C(\lambda)|p^\varepsilon(t) - p^*(t)|_{V^*}.$$

We take squares in both sides of the inequality, integrate over $[0,T]$ and obtain $p^\varepsilon \to p^*$ strongly in $L^2(Q)$.

Remark 2.5. Results for parabolic control problems, similar to Theorem 2.4, are given in the book of V.Barbu [13], Ch.V (see also its references). The proof we present here is simpler and it may also be applied to semilinear elliptic problems.

In the work [134] of F.Troltzsch, state constrained control problems, governed by semilinear parabolic equations, are studied under differentiability assumptions.

Remark 2.6. If no constraints are imposed on u, and L is quadratic, then, for $W = L^2(\Omega)$ (distributed control), by Thm.2.4, we see that $u^* \in L^2(0,T;H^1(\Omega))$. This regularity result also extends to some situations when control constraints are present. For instance, if we ask that $u(t) \in U_{ad}$ a.e. [0,T] with

$$U_{ad} = \{u \in L^2(\Omega);\ u \geq 0 \text{ a.e.} \Omega\},$$

then, by the maximum principle and the definition of the subdifferential, an elementary calculation gives $u^* = (F^* p^*)^-$, so again $u^* \in L^2(0,T;H^1(\Omega))$. Here $(f)^-$ is the negative part of f.

Remark 2.7. If $W = L^2(\Omega)$ and $u^* \in L^p(\Omega)$, $p \geq 2$, $p > (N+2)/2$, then we can use a variant of the above method and penalize by $1/p|u - u^*|^p_{L^p(Q)}$ in the approximate problem. It yields that $\{y^\varepsilon\}$ is bounded in $L^p(0,T;W^{2,p}(\Omega) \cap W_0^{1,p}(\Omega))$ and $\{y_t^\varepsilon\}$ is bounded in $L^p(Q)$ for sufficiently regular initial data, according to Ch.I, Thm.4.5..The Sobolev embedding theorem implies $y^\varepsilon \to y^*$ strongly in $C(\overline{Q})$ and it is sufficient to assume only that β is locally Lipschitzian, without the growth condition (2.5).

This situation appears, for instance, if the set of admissible controls is contained in $L^p(Q)$ or if L is coercive on $L^p(Q)$.

3. Hyperbolic problems

We start with the description of a control problem governed by a nonlinear second order, abstract evolution equation:

(3.1) $\quad$ Minimize $\int_0^T L(y,u)dt$,

(3.2) $\quad$ $y''(t) + Ay(t) + \partial\varphi(y'(t)) \ni Bu(t)$ $\quad$ a.e. $]0,T[$,

(3.3) $\quad$ $y(0) = y_0,\ y'(0) = v_0$.

Here we consider V, U, H Hilbert spaces, with $V \subset H \subset V^*$ continuously and densely, and $A : V \to V^*$ a linear, continuous symmetric operator such that:

(3.4) $\quad$ $(Av,v)_H + \sigma|v|^2_H \geq \omega|v|^2_V,\ \omega > 0,\ \sigma \in R.$

The mapping $\varphi : V \to]-\infty, +\infty]$ is convex, lower semicontinuous, proper, and $y_0 \in V$, $Ay_0 \in H$, $v_0 \in \mathrm{dom}(\varphi)$, $B : U \to V$ is a linear, bounded operator.

Existence and regularity results for (3.2), (3.3) may be found in Brezis [20], Barbu [12]. In particular, the unique solution y of (3.2), (3.3) satisfies $y \in C(0,T;V)$, $y' \in C(0,T;H) \cap L^\infty(0,T;V)$, $y'' \in L^2(0,T;H)$, $Ay \in L^\infty(0,T;H)$ if $u \in L^2(0,T;U)$.

We define the spaces $W = U$, $Z = V \times H$, $X = H$, $Y = \{0\} \times V$ and the operators

(3.5) $\quad$ $F : W \to Y,\quad F = [0,B]$,

(3.6) $\quad$ $S : Z \to X,\quad S(v,h) = v,\quad v \in V,\ h \in H$,

(3.7) $\quad$ $M : Z \to Z\quad M(v,h) = [-h, \{Av + \partial\varphi(h)\} \cap H] + \alpha[v,h]$, where $h \in dom(\partial\varphi)$ and α is given by:

$$\alpha = \sup\{\sigma(u,v)_H/((Au,u)_{V \times V^*} + \sigma|u|_H^2 + |v|_H^2) \; ; u \in V, \, v \in H; \, |u|_V + |v|_H \neq 0\} < \infty,$$

Here, we identify Z with its dual by endowing it with the inner product

$$([v_1,h_1],[v_2,h_2])_Z = (Av_1,v_2)_{V \times V^*} + (v_1,v_2)_V + (h_1,h_2)_H.$$

If the equation (3.2) is written as a first order system and we use the above notations, we obtain a problem of the type (1.1)-(1.3). The pair $[y,y']$ plays the role of the state from (1.2), (1.3).

We remark that the maximal monotone operator M, given by (3.7), isn't a subdifferential, in general. Therefore the situation investigated in this section is essentially different from § 2.

We continue with the study of two important problems of the type (3.1)-(3.3), by means of the scheme developed in § 1. Another example will be discussed in Ch.III, § 2.

3.1. Semilinear, second order hyperbolic equations

Let $\beta \subset R \times R$ be a maximal monotone graph, $0 \in \beta(0)$ and $j : R \to]-\infty,+\infty]$ be a convex, lower semicontinuous function, such that $\beta = \partial j$. We take $V = H_0^1(\Omega)$, $H = L^2(\Omega)$ and $\varphi : L^2(\Omega) \to]-\infty,+\infty]$ is given by

$$\varphi(y) = \begin{cases} \int_\Omega j(y(x))dx, & \text{if } y \in L^2(\Omega),\ j(y) \in L^1(\Omega), \\ +\infty & \text{otherwise}. \end{cases} \tag{3.8}$$

The mapping φ is convex, lower semicontinuous, proper on H and $\partial\varphi(y)(x) \in \beta(y(x))$ a.e. Ω for $y \in dom(\partial\varphi)$ (Ch.I, § 3). We define φ^ε by (3.8) with j^ε in the place of j, where j^ε is given by $j^\varepsilon(r) = \int_0^r [\tilde{\beta}^\varepsilon(s) - \tilde{\beta}^\varepsilon(0)]ds$ and $\tilde{\beta}^\varepsilon$ is defined by (2.8). This ensures that $0 = \beta^\varepsilon(0)$, for $\beta^\varepsilon = \partial j^\varepsilon$. The operator M^ε is obtained by (3.7), when φ is replaced by φ^ε.

We put $A : H_0^1(\Omega) \to H^{-1}(\Omega)$ to be the Laplace operator with zero Dirichlet conditions on the boundary, and $B : U \to H_0^1(\Omega)$ is linear, bounded.

Then, the equation (3.2), (3.3) models the movement of a membrane clamped at the border, with resistance proportional to the velocity.

The approximate control problem may be written as follows:

$$\text{Minimize}\left\{\int_0^T L^\varepsilon(y(t),u(t))dt + 1/2\int_0^T |u - u^*|_U^2 dt\right\}, \tag{3.9}$$

subject to

$$y_{tt} - \Delta y + \beta^\varepsilon(y_t) = Bu \qquad \text{a.e. } Q, \tag{3.10}$$

(3.11) $y(0,x) = y_o(x), \quad y_t(0,x) = v_o(x)$ a.e.Ω ,

(3.12) $y(t,x) = 0$ a.e. Σ .

Remark 3.1. According to the discussion from the beginning of this section θ_ε (and θ) is defined by a first order, nonlinear evolution system. We see that, by (3.5), F = [0,B] and it yields that the system may be always put in the form of a second order equation and θ_ε is equivalently defined by (3.10)-(3.12). We may take θ_ε and θ as defined on $L^2(0,T;H_o^1(\Omega))$.

Lemma 3.2. Let $u_\varepsilon \to u$ weakly in $L^2(0,T;U)$. Then $y^\varepsilon = (S \circ \theta_\varepsilon)(Bu_\varepsilon) \to y = (S \circ \theta)(Bu)$ strongly in $C(0,T;H_o^1(\Omega))$ and $y_t^\varepsilon \to y_t$ strongly in $C(0,T;L^2(\Omega))$.

Proof

We multiply (3.10) by y_t^ε and integrate over [0,t]:

$$1/2(\Delta y_o,y_o)_{L^2(\Omega)} + 1/2|y_t^\varepsilon(t)|^2_{L^2(\Omega)} - 1/2|v_o|^2_{L^2(\Omega)} + \omega/2|\mathrm{grad}\, y^\varepsilon(t)|^2_{L^2(\Omega)} \le$$

$$\le \int_0^t (Bu_\varepsilon, y_t^\varepsilon)_{L^2(\Omega)} ds.$$

Obviously $\beta^\varepsilon(y_t^\varepsilon) \in L^\infty(0,T;H_o^1(\Omega))$ and we have the inequality:

$$-\int_\Omega \Delta y_t^\varepsilon \beta^\varepsilon(y_t^\varepsilon)dx = \int_\Omega \mathrm{grad}\, y_t^\varepsilon \cdot \mathrm{grad}\, \beta^\varepsilon(y_t^\varepsilon)dx \ge 0,$$

by the monotonicity of β^ε and by $\beta^\varepsilon(0) = 0$.

Multiply (3.10) by Δy_t^ε and integrate over [0,t]:

$$1/2|y_t^\varepsilon(t)|^2_{H_o^1(\Omega)} - 1/2|v_o|^2_{H_o^1(\Omega)} + 1/2|\Delta y^\varepsilon(t)|^2_{L^2(\Omega)} - 1/2|\Delta y_o|^2_{L^2(\Omega)} \le$$

$$\le \int_0^t (Bu_\varepsilon, \Delta y_t^\varepsilon)_{H_o^1(\Omega) \times H^{-1}(\Omega)} ds$$

The above operations have a formal character, but they may be entirely justified by working with the Yosida approximation of the realization of $-\Delta$ in $L^2(\Omega)$ and next passing to the limit.

We integrate by parts with respect to x in the right-hand side and it yields that $\{y_t^\varepsilon\}$ is bounded in $L^\infty(0,T;H_o^1(\Omega))$, $\{\Delta y^\varepsilon\}$ is bounded in $L^\infty(0,T;L^2(\Omega))$ and $\{y^\varepsilon\}$ is bounded in $L^\infty(0,T;H^2(\Omega))$.

Now, we multiply (3.10) by y_{tt}^ε and integrate on [0,T]:

$$\int_0^T |y_{tt}^\varepsilon|^2_{L^2(\Omega)} + \varphi^\varepsilon(y_t^\varepsilon(T)) - \varphi^\varepsilon(v_o) \le C(|y_{tt}^\varepsilon|_{L^2(Q)} + 1)$$

since all the other terms are bounded. But, $0 = \beta^\varepsilon(0)$ implies that j^ε and φ^ε are positive. Moreover $v_o \in \mathrm{dom}(\varphi)$, so $\{\varphi^\varepsilon(v_o)\}$ is bounded with respect to $\varepsilon > 0$. It yields that $\{y_{tt}^\varepsilon\}$ is bounded in $L^2(Q)$ and, by (3.10), $\{\beta^\varepsilon(y_t^\varepsilon)\}$ is bounded in $L^2(Q)$ too.

The Arzela-Ascoli criterion gives:

$y^\varepsilon \to \tilde{y}$ strongly in $C(0,T;H_0^1(\Omega))$,
$y_t^\varepsilon \to \tilde{y}_t$ strongly in $C(0,T;L^2(\Omega))$,
$y_{tt}^\varepsilon \to \tilde{y}_{tt}$ weakly in $L^2(Q)$,
$\beta^\varepsilon(y_t^\varepsilon) \to w$ weakly in $L^2(Q)$

and $w \in \beta(y_t)$ a.e. Q due to the demiclosedness of maximal monotone operators. We can pass to the limit in (3.10) and obtain $\tilde{y} = \theta(Bu)$. By the uniqueness of $\theta(Bu)$, we see that the convergence is true on the initial sequence.

Remark 3.3. Lemma 3.2. checks the condition (a) of § 1 directly in the form it is used in the proofs of § 1.

We continue with hypotheses (b) and (c).

Lemma 3.4. The operator $S \circ \theta_\varepsilon : L^2(0,T;H_0^1(\Omega)) \to L^2(Q)$ is Gateaux differentiable and $r = \nabla(S \circ \theta_\varepsilon)(f)g$ satisfies:

(3.13) $\quad r_{tt} - \Delta r + \dot{\beta}^\varepsilon(y_t)r_t = g \quad$ in Q,
(3.14) $\quad r(0,x) = r_t(0,x) = 0 \quad$ in Ω,
(3.15) $\quad r(t,x) = 0 \quad$ in Σ,

where $y = (S \circ \theta_\varepsilon)(f)$ and $f, g \in L^2(0,T;H_0^1(\Omega))$. Moreover $r \in L^\infty(0,T;H_0^1(\Omega))$, $r_t \in L^\infty(0,T;L^2(\Omega))$, $r_{tt} \in L^2(0,T;H^{-1}(\Omega))$. The solution of (3.13)-(3.15) is unique.

Proof

Let $y^\lambda = (S \circ \theta_\varepsilon)(f + \lambda g)$. We have

(3.16) $\quad y_{tt} - \Delta y + \beta^\varepsilon(y_t) = f \quad$ a.e.Q,
(3.17) $\quad y_{tt}^\lambda - \Delta y^\lambda + \beta^\varepsilon(y_t^\lambda) = f + \lambda g \quad$ a.e.Q,
$\quad y(0) = y^\lambda(0) = y_0,\ y_t(0) = y_t^\lambda(0) = v_0 \quad$ a.e.Ω .

We subtract (3.16), (3.17) and multiply by $y_t^\lambda - y_t$:

(3.18) $$1/2\,|y_t^\lambda(t) - y_t(t)|^2_{L^2(\Omega)} + 1/2\,|y^\lambda(t) - y(t)|^2_{H_0^1(\Omega)} \le \lambda\int_0^t (g(s), y_t^\lambda(s) - y_t(s))_{L^2(\Omega)} ds$$

Then $y_t^\lambda \to y_t$ strongly in $C(0,T;L^2(\Omega))$, $y^\lambda \to y$ strongly in $C(0,T;H_0^1(\Omega))$ and $\{z^\lambda\}$, $\{z_t^\lambda\}$ are bounded in $L^\infty(0,T;H_0^1(\Omega))$, respectively $L^\infty(0,T;L^2(\Omega))$ where $z^\lambda = (y^\lambda - y)/\lambda$.

It yields that $z^\lambda \to r$ strongly in $C(0,T;L^2(\Omega))$, $z_t^\lambda \to r_t$ weakly* in $L^\infty(0,T;L^2(\Omega))$.

On the other side, we have

$$(\beta^\varepsilon(y_t^\lambda) - \beta^\varepsilon(y_t))/\lambda = (\beta^\varepsilon(y_t^\lambda) - \beta^\varepsilon(y_t))z_t^\lambda/(y_t^\lambda - y_t) \qquad \text{a.e.}Q$$

and, from the above convergences, as well as from the properties of β^ε, we get

$$(\beta^\varepsilon(y_t^\lambda) - \beta^\varepsilon(y_t))/\lambda \to \dot\beta^\varepsilon(y_t)r_t$$

weakly in $L^2(Q)$.

Lemma 3.5. Let $y^\varepsilon = (S \circ \theta_\varepsilon)(f)$ and $y = (S \circ \theta)(f)$, $f \in L^2(0,T;H_0^1(\Omega))$; then:

$$|y^\varepsilon(t) - y(t)|_{H_0^1(\Omega)} + |y_t^\varepsilon(t) - y_t(t)|_{L^2(\Omega)} \le C\varepsilon^{1/2}, \quad t \in [0,T],$$

where C is a constant which depends on $|f|_{L^2(0,T;H_0^1(\Omega))}, |y_0|_{H_0^1(\Omega)}, |v_0|_{L^2(\Omega)}$.

Proof

Subtract the equations corresponding to θ_ε, θ_λ and multiply by $y_t^\varepsilon - y_t^\lambda$:

$$\frac{1}{2}|y_t^\varepsilon(t) - y_t^\lambda(t)|^2_{L^2(\Omega)} + \frac{1}{2}|y^\varepsilon(t) - y^\lambda(t)|^2_{H_0^1(\Omega)} + $$
$$+ \int_o^t (\beta^\varepsilon(y_t^\varepsilon) - \beta^\lambda(y_t^\lambda), y_t^\varepsilon - y_t^\lambda)_{L^2(\Omega)} dt = 0 .$$

By a computation similar to the proof of Lemma 2.2 and by the boundedness of $\{\beta^\varepsilon(y_t^\varepsilon)\}$ in $L^2(Q)$, we get

$$\frac{1}{2}|y_t^\varepsilon(t) - y_t^\lambda(t)|^2_{L^2(\Omega)} + \frac{1}{2}|y^\varepsilon(t) - y^\lambda(t)|^2_{H_0^1(\Omega)} \le C(\varepsilon + \lambda).$$

Using the convergence obtained in Lemma 3.2 we can pass to the limit $\lambda \to 0$ and conclude the proof.

Proposition 3.6. The approximate problem (3.9) - (3.12) has at least one optimal solution $[y^\varepsilon, u_\varepsilon]$ in $W^{2,2}(0,T;L^2(\Omega)) \times L^2(0,T;U)$ and there is $m^\varepsilon \in C(0,T;L^2(\Omega))$ such that, the approximate optimality system

(3.19) $\quad m_{tt}^\varepsilon - \Delta m^\varepsilon - \dot\beta^\varepsilon(y_t^\varepsilon)m_t^\varepsilon = -\int_t^T q_\varepsilon \quad$ in Q,

(3.20) $\quad m^\varepsilon(T,x) = 0,\ m_t^\varepsilon(T,x) = 0 \quad$ in Ω,

(3.21) $\quad m^\varepsilon(t,x) = 0 \quad$ in Σ,

(3.22) $\quad [q_\varepsilon(t), -B^* m_t(t) - u_\varepsilon(t) + u^*(t)] = \partial L^\varepsilon(y^\varepsilon(t), u_\varepsilon(t))$ in $[0,T]$

is satisfied together with (3.10) - (3.12). Furthermore $y^\varepsilon \to y^*$ strongly in $C(0,T;H_0^1(\Omega))$, $y_t^\varepsilon \to y_t^*$ strongly in $C(0,T;L^2(\Omega))$, $u_\varepsilon \to u^*$ strongly in $L^2(0,T;U)$, $p^\varepsilon = -m_t^\varepsilon \to p^*$ weakly* in $L^\infty(0,T;L^2(\Omega))$ and $q_\varepsilon \to q^*$ weakly in $L^1(0,T;L^2(\Omega))$, where

(3.23) $\quad [q^*(t), B^* p^*(t)] \in \partial L(y^*(t), u^*(t)) \quad$ in $[0,T]$

and $[y^*, u^*]$ is an optimal pair for (3.1)-(3.3).

Proof

The system (3.19)-(3.21) has a unique solution $m^\varepsilon \in L^\infty(0,T;H_0^1(\Omega)) \cap W^{1,\infty}(0,T;L^2(\Omega))$, according to Lemma 3.7 below. The fact that $p^{\varepsilon} = -m_t^{\varepsilon} = -\nabla(S \circ \theta_{\varepsilon})(Bu_{\varepsilon})^*(q_{\varepsilon})$ is a standard consequence of Lemma 3.4 and some integration by parts, according to the definition of the adjoint operator.

Lemma 3.7. The problem (3.19)-(3.21) has a unique solution $m^{\varepsilon} \in L^{\infty}(0,T;H_0^1(\Omega)) \cap W^{1,\infty}(0,T;L^2(\Omega))$, in the distribution sense.

Proof

We reverse the sense of the time and write (3.19)-(3.21) as a first order evolution system by means of the operator

$$P = \begin{bmatrix} 0 & -1 \\ -\Delta & \dot{\beta}^{\varepsilon}(y_t) \end{bmatrix}$$

with the domain and the range included in $L^2(\Omega)^2$. We approximate P by the operator

$$P^{\lambda} = \begin{bmatrix} \lambda & -1 \\ A_{\lambda} & \lambda + \dot{\beta}^{\varepsilon}(y_t) \end{bmatrix},$$

$\lambda > 0$. Here A_{λ} is the Yosida approximation of the realization A_H of $-\Delta$ in $L^2(\Omega)$.

On $L^2(\Omega)^2$ we take the equivalent scalar product, denoted "$\cdot$"

$$[y_1,v_1] \cdot [y_2,v_2] = (A_{\lambda}y_1,y_2) + (y_1,y_2) + 1/\lambda(v_1,v_2),$$

where (.,.) is the inner product in $L^2(\Omega)$.

The operator $P^{\lambda}(t)$, $t \in [0,T]$ is linear, continuous, coercive on $L^2(\Omega)^2$:

$$P^{\lambda}(t)([y,v]).[y,v] = (-A_{\lambda}v,y) + \lambda(A_{\lambda}y,y) - \lambda(v,y) + \lambda^2|y|^2_{L^2(\Omega)} + 1/\lambda(A_{\lambda}y,v) +$$

$$+ |v|^2_{L^2(\Omega)} + 1/\lambda(\dot{\beta}^{\varepsilon}(y_t^{\varepsilon})v,v) \geq \lambda^2|y|^2_{L^2(\Omega)} + |v|^2_{L^2(\Omega)} - \lambda(v,y).$$

We may apply Thm.4.4, Ch.I, and obtain the existence of a solution $[y^{\lambda},v^{\lambda}] \in C(0,T;L^2(\Omega)^2)$, $[y_t^{\lambda},v_t^{\lambda}] \in L^2(0,T;L^2(\Omega)^2)$, satisfying the evolution system:

(3.24) $\quad y_t^{\lambda} - v^{\lambda} + \lambda y^{\lambda} = 0 \quad$ in Q,

(3.25) $\quad v_t^{\lambda} + A_{\lambda}y^{\lambda} + \lambda v^{\lambda} + \dot{\beta}^{\varepsilon}(y_t^{\varepsilon})v^{\lambda} = f \quad$ in Q,

$\quad y^{\lambda}(0) = y_0, \; v^{\lambda}(0) = v_0 \quad$ in Ω,

for any $f \in L^2(Q)$.

Multiply by v^{λ} in (3.25) and integrate over $[0,t]$:

$$1/2|v^\lambda(t)|^2_{L^2(\Omega)} - 1/2|v_o|^2_{L^2(\Omega)} + \int_o^t (A_\lambda y^\lambda, y_t^\lambda + \lambda y^\lambda) ds \leq \int_o^t (f, v^\lambda).$$

It yields

$$1/2(A_\lambda y^\lambda(t), y^\lambda(t)) + 1/2|v^\lambda(t)|^2_{L^2(\Omega)} \leq C + \int_o^t (f, v^\lambda).$$

Therefore $\{v^\lambda\}$ is bounded in $L^\infty(0,T;L^2(\Omega))$ and $\{(I + \lambda A_H)^{-1} y^\lambda\}$ is bounded in $L^\infty(0,T;H_o^1(\Omega))$ because $(A_\lambda y, y) \geq |(I + \lambda A_H)^{-1} y|^2_{H_o^1(\Omega)}$.

We multiply by y^λ in (3.24) and we deduce that $\{y^\lambda\}$ is bounded in $L^\infty(0,T;L^2(\Omega))$, $\{y_t^\lambda\}$ is bounded in $L^\infty(0,T;L^2(\Omega))$.

By the definition, we have that $\{A_\lambda y^\lambda\}$ is bounded in $L^\infty(0,T;H^{-1}(\Omega))$ and (3.25) implies that $\{v_t^\lambda\}$ is bounded in $L^2(0,T;H^{-1}(\Omega))$.

But $A_\lambda y^\lambda = \lambda^{-1}(y^\lambda - (I + \lambda A_H)^{-1} y^\lambda)$ and so $y^\lambda - (I + \lambda A)^{-1} y^\lambda \to 0$ strongly in $L^\infty(0,T;H^{-1}(\Omega))$. We have shown that there exist $\tilde{y} \in L^\infty(0,T;H_o^1(\Omega))$, $\tilde{v} \in L^\infty(0,T;L^2(\Omega))$, such that

$$\begin{aligned}
&(I + \lambda A_H)^{-1} y^\lambda \to \tilde{y} \text{ weakly* in } L^\infty(0,T;H_o^1(\Omega)),\\
&y^\lambda \to \tilde{y} \text{ weakly* in } L^\infty(0,T;L^2(\Omega)),\\
&v^\lambda \to \tilde{v} \text{ weakly* in } L^\infty(0,T;L^2(\Omega)),\\
&y_t^\lambda \to \tilde{y}_t \text{ weakly* in } L^\infty(0,T;L^2(\Omega)),\\
&v_t^\lambda \to \tilde{v}_t \text{ weakly in } L^2(0,T;H^{-1}(\Omega)),\\
&A_\lambda y^\lambda \to -\Delta\tilde{y} \text{ weakly* in } L^\infty(0,T;H^{-1}(\Omega)).
\end{aligned}$$

We conclude that $\tilde{y} \in L^\infty(0,T;H_o^1(\Omega))$ and $\tilde{v} \in L^\infty(0,T;L^2(\Omega))$ satisfy the system:

$$\begin{aligned}
&\tilde{y}_t - \tilde{v} = 0\\
&\tilde{v}_t - \Delta\tilde{y} + \dot{\beta}^\varepsilon(y^\varepsilon)\tilde{v} = f,\\
&\tilde{y}(0) = y_o,\ \tilde{v}(0) = v_o.
\end{aligned}$$

Therefore, when $f = -\int_{T-t}^T q_\varepsilon \, ds$ and the time is again reversed, we obtain the desired result.

To pass to the limit in (3.19) - (3.21), we impose again hypothesis (2.5). The identification of the limit of the product $\dot{\beta}^\varepsilon(y_t^\varepsilon) m_t^\varepsilon$ is very difficult since both factors are only weakly convergent sequences.

Theorem 3.8. Let $[y^*, u^*] \in W^{2,2}(0,T;L^2(\Omega)) \times L^2(0,T;U)$ be an optimal pair for the problem (3.1) - (3.3) and β satisfy (2.5). There are $m^* \in L^\infty(0,T;H_o^1(\Omega)) \cap W^{1,\infty}(0,T;L^2(\Omega))$, $q^* \in L^2(Q)$ and $h \in L^1(Q)$ such that:

$$\begin{aligned}
&m_{tt}^* - \Delta m^* - h = -\int_t^T q^* && \text{in } Q,\\
&m^*(T,x) = m_t^*(T,x) = 0 && \text{in } \Omega .
\end{aligned}$$

Moreover y^*, u^*, m^*, q^* are limits of the solutions $y^\varepsilon, u_\varepsilon$, m^ε, q_ε given by Proposition 3.6.

Proof

We multiply (3.19) by m_t^ε and integrate over [t,T]. Since $\{q_\varepsilon\}$ is bounded in $L^1(0,T;L^2(\Omega))$ and $\dot\beta^\varepsilon$ is positive, we obtain $\{m^\varepsilon\}$ bounded in $L^\infty(0,T;H_0^1(\Omega))$, $\{m_t^\varepsilon\}$ bounded in $L^\infty(0,T;L^2(\Omega))$ and $\{\dot\beta^\varepsilon(y_t^\varepsilon)|m_t^\varepsilon|^2\}$ bounded in $L^1(Q)$.

Condition (2.5) implies directly that $\{\dot\beta^\varepsilon(y_t^\varepsilon)\}$ is bounded in $L^2(Q)$ because $\{\beta^\varepsilon(y_t^\varepsilon)\}$ is so.

We use a trick based on the inequality of Young to show that $\{\dot\beta^\varepsilon(y_t^\varepsilon)m_t^\varepsilon\}$ is bounded in $L^s(Q)$, for some s > 1:

$$|\dot\beta^\varepsilon(y_t^\varepsilon)m_t^\varepsilon| = [\dot\beta^\varepsilon(y_t^\varepsilon)]^{1/2}[(\dot\beta^\varepsilon(y_t^\varepsilon))^{1/2}|m_t^\varepsilon|] \le 1/3[\dot\beta^\varepsilon(y_t^\varepsilon)]^{3/2} + 2/3[\dot\beta^\varepsilon(y_t^\varepsilon)|m_t^\varepsilon|^2]^{3/4}.$$

This inequality and the above estimates show that we can choose, for instance, s = 4/3 and $\{\dot\beta^\varepsilon(y_t^\varepsilon)m_t^\varepsilon\}$ is bounded in $L^{4/3}(Q)$.

On a subsequence, we have $\dot\beta^\varepsilon(y_t^\varepsilon)m_t^\varepsilon \to h$ weakly in $L^{4/3}(Q)$ and we can pass to the limit in (3.19)-(3.21).

Remark 3.9. By the theorem of Egorov, for any $\eta > 0$, there is $Q_\eta \subset Q$, mes$(Q - Q_\eta) < \eta$ and $y_t^\varepsilon \to y_t^*$ uniformly on Q_η. As β is locally Lipschitzian, we see that

$$|\dot\beta^\varepsilon(y_t^\varepsilon)| \le C \qquad \text{a.e. } Q$$

and Thm.3.14, Ch.I, implies that

$$\dot\beta^\varepsilon(y_t^\varepsilon) \to \gamma^* \in D\beta(y_t^*)$$

weakly* in $L^\infty(Q_\eta)$. One may expect to obtain the relation

$$h(t,x) \in D\beta(y_t^*(t,x))m_t^*(t,x) \qquad \text{a.e.}Q. \tag{3.26}$$

If β is of class $C^1(R)$, then

$$\dot\beta^\varepsilon(y_t^\varepsilon) \to \dot\beta(y_t^*) \qquad \text{a.e.}Q$$

and the Lebesgue theorem gives $\dot\beta^\varepsilon(y_t^\varepsilon) \to \dot\beta(y_t^*)$ strongly in $L^2(Q_\eta)$. It yields $\dot\beta^\varepsilon(y_t^\varepsilon)m_t^\varepsilon \to \dot\beta(y_t^*)m_t^*$ weakly in $L^1(Q_\eta)$ and (3.26) is checked. (Obviously, the case $\beta \in C^1(R)$ may be discussed by a more direct argument).

We continue by establishing relation (3.26) under certain convexity assumptions:

$$\beta = \alpha - \delta, \tag{3.27}$$

where α, δ are real convex functions, finite on the whole real axis. In particular, it yields that β is locally Lipschitzian.

Theorem 3.10. Under hypotheses (2.5), (3.27) there exist the functions $m^* \in L^\infty(0,T;H_0^1(\Omega)) \cap W^{1,\infty}(0,T;L^2(\Omega))$ and $q^* \in L^2(Q)$ such that we have

$$\begin{aligned} & m^*_{tt} - \Delta m^* - D\beta(y^*_t).m^*_t \ni - \int_t^T q^* && \text{in } Q, \\ & m^*(T,x) = m^*_t(T,x) = 0 && \text{in } \Omega, \\ & [q^*(t), - B^* m^*_t(t)] \in \partial L(y^*(t),u^*(t)) && \text{a.e. } [0,T], \end{aligned}$$

in a weak sense.

Proof

Let us consider first β convex. In order to identify the limit of the product $\dot\beta^\varepsilon(y^\varepsilon_t)m^\varepsilon_t$, we write $m^\varepsilon_t = m^\varepsilon_+ - m^\varepsilon_-$, where m^ε_+, m^ε_- are the positive and the negative parts of m^ε_t, up to a constant, such that both of them are strictly positive.

By taking further subsequences, if necessary, we get

$$\begin{aligned} & m^\varepsilon_+ \to m^*_+ && \text{weakly in } L^2(Q), \\ & m^\varepsilon_- \to m^*_- && \text{weakly in } L^2(Q), \\ (3.28) \qquad & m^*_t = m^*_+ - m^*_- \,. \end{aligned}$$

We indicate a precise computation of $\dot\beta^\varepsilon(y)$, valid for locally Lipschitzian mappings

$$(3.29) \qquad \beta^\varepsilon(y) = \int_{-\infty}^{\infty} (I - (I + \varepsilon\beta)^{-1})(y - \varepsilon\tau)\rho(\tau)/\varepsilon \, d\tau.$$

Since β is almost everywhere differentiable and $(I + \varepsilon\beta)^{\cdot} = 1 + \varepsilon\dot\beta \geq 1$ a.e., then $(I + \varepsilon\beta)^{-1}$ is differentiable a.e. and we obtain

$$(3.30) \qquad \dot\beta^\varepsilon(y) = \int_{-\infty}^{\infty} \dot\beta((I + \varepsilon\beta)^{-1}(y - \varepsilon\tau))/(1 + \varepsilon\dot\beta((I + \varepsilon\beta)^{-1}(y - \varepsilon\tau)))\rho(\tau)d\tau.$$

When β is convex, then $\dot\beta = \partial\beta$ (the subdifferential of β) in the points where β is differentiable, so

$$(3.31) \qquad \dot\beta^\varepsilon(y) = \int_{-1}^{1} \partial\beta((I + \varepsilon\beta)^{-1}(y - \varepsilon\tau))/(1 + \varepsilon\,\partial\beta((I + \varepsilon\beta)^{-1}(y - \varepsilon\tau)))\rho(\tau)d\tau.$$

Let Q_η be as in Remark 2.9. We study the weak convergence of $\dot\beta^\varepsilon(y^\varepsilon_t).m^\varepsilon_+$ in $L^2(Q_\eta)$. Let $f \in L^2(Q_\eta)$ be arbitrary fixed. Then:

$$\int_{Q_\eta} \dot\beta^\varepsilon(y^\varepsilon_t).m^\varepsilon_+.f dx dt = \int_{-1}^{1} \rho(\tau)d\tau \int_{Q_\eta} m^\varepsilon_+ \partial\beta((I + \varepsilon\beta)^{-1}(y_t - \varepsilon\tau))f/(1 + \varepsilon\,\partial\beta(.))dx dt.$$

Because $y^\varepsilon_t - \varepsilon\tau$ is uniformly bounded on Q_η, for $\varepsilon \to 0$, then $\partial\beta((I + \varepsilon\beta)^{-1}(y^\varepsilon_t - \varepsilon\tau))$ is bounded in $L^\infty(Q_\eta)$ and the last factor converges uniformly to 1 on Q_η. Therefore, we have to analyse the integral

(3.32) $$\int_{Q_\eta} m_+^\varepsilon \partial\beta((I+\varepsilon\beta)^{-1}(y_t^\varepsilon-\varepsilon\tau))f\,dxdt, \qquad \tau\in[-1,1].$$

We define the concave-convex function

(3.33) $$K(m,y) = \begin{cases} m\beta(y) & m \geq 0, \\ -\infty, & m < 0, \end{cases}$$

which is proper, closed, according to § 3, Ch.I.

The maximal monotone operator $\partial K \subset R^2 \times R^2$ satisfies

(3.34) $$[-\beta(y),\ m\partial\beta(y)] \in \partial K(m,y), \quad \forall\ [m,y]\in \mathrm{dom}(\partial K).$$

We consider the realization of ∂K in $L^2(Q_\eta)\times L^2(Q_\eta)$:

(3.35) $$\partial\tilde{K}(m,y)(t,x) \in \partial K(m(t,x),y(t,x)) \qquad \text{a.e.}Q\ ,$$

for $m,y\in L^2(Q_\eta)$, $m \geq 0$, a.e. Q_η.

The operator $\partial\tilde{K}$ is maximal monotone in $L^2(Q_\eta)\times L^2(Q_\eta)$ and

(3.36) $$[-\beta((I+\varepsilon\beta)^{-1}(y_t^\varepsilon-\varepsilon\tau)),\ m_+^\varepsilon\partial\beta((I+\varepsilon\beta)^{-1}(y_t^\varepsilon-\varepsilon\tau))] \in \partial\tilde{K}(m_+^\varepsilon,\ (I+\varepsilon\beta)^{-1}(y_t^\varepsilon-\varepsilon\tau)).$$

We remark the convergences:

(3.37) $$[-\beta((I+\varepsilon\beta)^{-1}(y_t^\varepsilon-\varepsilon\tau)),\ m_+^\varepsilon\partial\beta((I+\varepsilon\beta)^{-1}(y_t^\varepsilon-\varepsilon\tau))] \to [-\beta(y_t^*),\tilde{h}],$$

(3.38) $$[m_+^\varepsilon,(I+\varepsilon\beta)^{-1}(y_t^\varepsilon-\varepsilon\tau)] \to [m_+^*,\ y_+^*]$$

weakly in $L^2(Q_\eta)\times L^2(Q_\eta)$.

We can also check the following condition

(3.39) $$\lim_{\lambda,\varepsilon\to 0} < [m_+^\varepsilon,(I+\varepsilon\beta)^{-1}(y_t^\varepsilon-\varepsilon\tau)] - [m_+^\lambda,\ (I+\lambda\beta)^{-1}(y_t^\lambda-\lambda\tau)]\ ,\ [-\beta((I+\varepsilon\beta)^{-1}(y_t^\varepsilon-\varepsilon\tau),$$
$$m_+^\varepsilon\partial\beta((I+\varepsilon\beta)^{-1}(y_t^\varepsilon-\varepsilon\tau))] - [-\beta(I+\lambda\beta)^{-1}(y_t^\lambda-\lambda\tau)),m_+^\lambda\partial\beta((I+\lambda\beta)^{-1}(y_t^\lambda-\lambda\tau))] >$$
$$>_{L^2(Q_\eta)\times L^2(Q_\eta)} = 0,$$

because $(I+\varepsilon\beta)^{-1}(y_t^\varepsilon-\varepsilon\tau)\to y_t^*$ strongly in $L^2(Q_\eta)$, for $\varepsilon\to 0$.

By means of Thm.2.9, Ch.I, from (3.36)-(3.39), we conclude

(3.40) $$[-\beta(y_t^*),\ \tilde{h}] \in \partial K(m_+^*,\ y_t^*)$$

(3.41) $$\tilde{h}(t,x) \in m_+^*\partial\beta(y_t^*\ (t,x)) \qquad \text{a.e } Q.$$

Similarly, one may prove that

(3.42) $$\lim_{\varepsilon\to 0} m_-^{\varepsilon}\partial\beta((I+\varepsilon\beta)^{-1}(y_t^{\varepsilon}-\varepsilon\tau)) = \underset{\sim}{h}$$

weakly in $L^2(Q_\eta)$ and

(3.43) $$\underset{\sim}{h}(t,x)\in m_-^*\partial\beta(y_t^*(t,x)) \quad \text{a.e.} Q.$$

However, the sections of $\partial\beta(y_t^*)$ for which (3.41), (3.43) are satisfied, may be different, although $\partial\beta(.)$ is almost everywhere single-valued. Consequently, we may write

(3.44) $$h(t,x) = \tilde{h}(t,x) - \underset{\sim}{h}(t,x)\in m_t^*(t,x)\partial\beta(y_t^*(t,x))$$

only by convention, the meaning of (3.44) being given by (3.41), (3.43).

Now, we return to the general hypothesis (3.27). We define

(3.45) $$\alpha^{\varepsilon}(y) = \int_{-\infty}^{\infty}\alpha((I+\varepsilon\beta)^{-1}(y-\varepsilon\tau))\rho(\tau)d\tau,$$

(3.46) $$\delta^{\varepsilon}(y) = \int_{-\infty}^{\infty}\delta((I+\varepsilon\beta)^{-1}(y-\varepsilon\tau))\rho(\tau)d\tau,$$

and we have $\beta^{\varepsilon}(y) = \alpha^{\varepsilon}(y) - \delta^{\varepsilon}(y)$, $\dot{\beta}^{\varepsilon}(y) = \dot{\alpha}^{\varepsilon}(y) - \dot{\delta}^{\varepsilon}(y)$. We compute as in (3.31):

(3.47) $$\dot{\alpha}^{\varepsilon}(y) = \int_{-\infty}^{\infty}\partial\alpha((I+\varepsilon\beta)^{-1}(y-\varepsilon\tau))/(1+\varepsilon\dot{\beta}((I+\varepsilon\beta)^{-1}(y-\varepsilon\tau)))\rho(\tau)d\tau,$$

where $\partial\alpha$ is the subdifferential of the convex function α.

Under the locally Lipschitzian assumption, we have:

$$\dot{\beta}^{\varepsilon}(y_t^{\varepsilon})\to\gamma^*\in D\beta(y_t^*) \quad \text{weakly* in } L^{\infty}(Q_\eta),$$
$$\dot{\alpha}^{\varepsilon}(y_t)\to v\in\partial\alpha(y_t^*) \quad \text{weakly* in } L^{\infty}(Q_\eta);$$
$$\dot{\delta}^{\varepsilon}(y_t)\to w\in\partial\delta(y_t^*) \quad \text{weakly* in } L^{\infty}(Q_\eta)$$

and $\gamma^* = v - w$. Using the same procedure as before, from (3.45), (3.46), we obtain

$$m_t^{\varepsilon}.\dot{\alpha}^{\varepsilon}(y_t^{\varepsilon})\to v\,.\,m_t^*,$$
$$m_t^{\varepsilon}.\dot{\delta}^{\varepsilon}(y_t^{\varepsilon})\to w\,.\,m_t^*.$$

Therefore $m_t^{\varepsilon}.\dot{\beta}^{\varepsilon}(y_t^{\varepsilon})\to\gamma^*\,.\,m_t^*\in m_t^*\,.\,D\beta(y_t^*)$, weakly in $L^1(Q)$, under the same convention as in the case β convex.

3.2. Nonlinear boundary conditions

We devote this section to the study of the problem

(3.48) Minimize $\int_0^T L(y,u)dt$,

(3.49) $y_{tt} - \Delta y = Bu$ a.e. Q,

(3.50) $y(0,x) = y_0(x),\ y_t(0,x) = v_0(x)$ a.e.Ω,

(3.51) $-\partial y/\partial n \in \beta(y_t)$ a.e.Σ.

Here, $\beta \subset R \times R$ is a maximal monotone graph and $y_0 \in H^2(\Omega)$, $v_0 \in H^1(\Omega)$ with the compatibility condition $-\partial y_0/\partial n \in \beta(v_0)$ a.e.Γ.

It is known that, for the right-hand side in $W^{1,1}(0,T;L^2(\Omega))$ or in $L^2(0,T;H_0^1(\Omega))$, (3.49)-(3.51) has a unique solution $y \in L^\infty(0,T;H^2(\Omega))$, $y_t \in L^\infty(0,T;H^1(\Omega))$, $y_{tt} \in L^2(0,T;L^2(\Omega))$ (see Brezis [20], Tiba [125]).

We take $V = H^1(\Omega)$, $H = L^2(\Omega)$, $A : V \to V^*$, $(Ay,v)_{V\times V^*} = \int_\Omega \text{grad}\, y \cdot \text{grad}\, v dx$, $B : U \to H_0^1(\Omega)$ linear, bounded, $\varphi : V \to]-\infty,+\infty]$ convex, lower semicontinuous, proper, given by

$$\varphi(y) = \begin{cases} \int_\Omega j(y(\sigma))d\sigma & \text{if } j(y) \in L^1(\partial\Omega), \\ +\infty & \text{otherwise,} \end{cases} \tag{3.52}$$

where $j : R \to]-\infty,+\infty]$ is choosen such that $\beta = \partial j$.

Using these notations, a simple calculus based on the Green formula, shows that the problem (3.48)-(3.51) may be put in the form (3.1)-(3.3) and enters in the setting discussed in § 1:

$$\int_\Omega (y_{tt} - Bu)(v - y_t)dx = \int_\Omega \Delta y(v - y_t)dx = -\int_\Omega \text{grad}\, y \cdot \text{grad}\,(v - y_t) +$$

$$+ \int_\Gamma \frac{\partial y}{\partial n}(v - y_t)d\sigma = -(Ay, v - y_t)_{V\times V^*} - \int_\Gamma \beta(y_t)(v - y_t)d\sigma \leq$$

$$\leq -(Ay, v - y_t)_{V\times V^*} + \varphi(v) - \varphi(y_t), \quad \forall\, v \in H^1(\Omega).$$

In this section, we make the supplementary assumption that β is strongly monotone, that is

(3.53) $\beta = aI + \alpha$,

$a > 0$ and $\alpha \subset R \times R$ is maximal monotone.

The regularization β^ε of β is defined as follows

(3.54) $\beta^\varepsilon(y) = ay + \alpha^\varepsilon(y)$,

where α^ε is given by (2.8).

Let $j^\varepsilon(y) = \int_0^y \beta^\varepsilon(s)ds$ and φ^ε be given by (3.52) when j is replaced by j^ε. The operator M^ε is again introduced by (3.7) with φ^ε instead of φ. Therefore, the approximate control problem is:

(3.55) Minimize $\left\{ \int_0^T L^\varepsilon(y,u)dt + 1/2\int_0^T |u - u^*|_U^2 dt \right\}$, subject to (3.49), (3.50) and

$$(3.56)\qquad -\partial y/\partial n = \beta^{\varepsilon}(y_t) \qquad \text{a.e.}\,\Sigma.$$

Lemma 3.9. *Let* $u_\varepsilon \to u$ *weakly in* $L^2(0,T;U)$. *Then* $y^\varepsilon = (S\circ\theta_\varepsilon)(Bu_\varepsilon) \to y = (S\circ\theta)(Bu)$ *strongly in* $C(0,T;H^1(\Omega))$, $y_t^\varepsilon \to y_t$ *strongly in* $C(0,T;L^2(\Omega))$.

Proof

y^ε is the solution of (3.49), (3.50), (3.56) corresponding to u_ε. We multiply (3.49) by y_t^ε and we deduce that $\{y^\varepsilon\}$ is bounded in $L^\infty(0,T;H^1(\Omega))$, $\{y_t^\varepsilon\}$ is bounded in $L^\infty(0,T;L^2(\Omega))$.

We integrate (3.49) over the interval $[t,t+s]$ and we divide by $s>0$. Next we multiply by $-\Delta(y^\varepsilon(t+s)-y^\varepsilon(t))/s$. We have:

$$1/2\, d/dt\,|\mathrm{grad}(y^\varepsilon(t+s)-y^\varepsilon(t))/s|^2_{L^2(\Omega)} + 1/2\, d/dt\,|\Delta 1/s\int_t^{t+s} y^\varepsilon|^2_{L^2(\Omega)} \le$$

$$\le -\int_\Omega 1/s\int_t^{t+s} Bu_\varepsilon\, d\tau\, \Delta(y^\varepsilon(t+s)-y^\varepsilon(t))/s\, dx.$$

Here, we use the inequality:

$$-\int_\Omega \Delta(y^\varepsilon(t+s)-y^\varepsilon(t))/s\,.(y_t^\varepsilon(t+s)-y_t^\varepsilon(y))/s\, dx =$$
$$= 1/2\, d/dt\,|\mathrm{grad}(y^\varepsilon(t+s)-y^\varepsilon(t))/s|^2_{L^2(\Omega)} - \int_\Gamma \partial/\partial n(y^\varepsilon(t+s)-y^\varepsilon(t))s\,.(y_t^\varepsilon(t+s)-y_t^\varepsilon(t))/s$$
$$= 1/2\, d/dt\,|\mathrm{grad}(y^\varepsilon(t+s)-y^\varepsilon(t))/s|^2_{L^2(\Omega)} + \int_\Gamma (\beta^\varepsilon(y_t^\varepsilon(t+s)) - \beta^\varepsilon(y_t^\varepsilon(t))/s\,.\,(y_t^\varepsilon(t+s)-y_t^\varepsilon(t))/s\,d\sigma$$
$$\ge 1/2\, d/dt\,|\mathrm{grad}(y^\varepsilon(t+s)-y^\varepsilon(t))/s|^2_{L^2(\Omega)}\,.$$

We integrate over $[0,t]$:

$$1/2\,|\mathrm{grad}(y^\varepsilon(t+s)-y^\varepsilon(t))/s|^2_{L^2(\Omega)} - 1/2\,|\mathrm{grad}(y^\varepsilon(s)-y^\varepsilon(0))/s|^2_{L^2(\Omega)} +$$
$$+ 1/2\,|\Delta 1/s\int_t^{t+s} y^\varepsilon d\tau|^2_{L^2(\Omega)} - 1/2\,|\Delta 1/s\int_0^s y^\varepsilon d\tau|^2_{L^2(\Omega)} \le$$
$$\le -\int_0^t (1/s\int_\tau^{\tau+s} Bu_\varepsilon)\Delta(y^\varepsilon(\tau+s)-y^\varepsilon(\tau))/s\, dx d\tau.$$

Let $s\to 0$, which is possible due to the regularity of y^ε:

$$1/2\,|\mathrm{grad}\, y_t^\varepsilon(t)|^2_{L^2(\Omega)} - 1/2\,|\mathrm{grad}\, v_0|^2_{L^2(\Omega)} + 1/2\,|\Delta y^\varepsilon(t)|^2_{L^2(\Omega)} - 1/2\,|\Delta y_0|^2_{L^2(\Omega)} \le$$
$$\le -\int_0^t \int_\Omega Bu_\varepsilon(\tau)\Delta y_t^\varepsilon(\tau)\, dx d\tau,$$

where, in the right-hand side, the integral $\int_\Omega$ is in fact the pairing between V and V^*.

Since Δ is a linear, continuous operator between V and V^* and $\{y_t^\varepsilon\}$ is bounded in $L^\infty(0,T;L^2(\Omega))$, we get

(3.57) $$1/2|\text{grad } y_t^\varepsilon(t)|^2_{L^2(\Omega)} - 1/2|\text{grad } v_o|^2_{L^2(\Omega)} + 1/2|\Delta y^\varepsilon(t)|^2_{L^2(\Omega)} - 1/2|\Delta y_o|^2_{L^2(\Omega)} \leq$$

$$\leq C + \int_0^t |\text{grad } y_t^\varepsilon|^2_{L^2(\Omega)}.$$

The Gronwall lemma shows that $\{y_t^\varepsilon\}$ is bounded in $L^\infty(0,T;H^1(\Omega))$, $\{\Delta y^\varepsilon\}$ is bounded in $L^\infty(0,T;L^2(\Omega))$ and $\{y_{tt}^\varepsilon\}$ is bounded in $L^2(0,T;L^2(\Omega))$. We prove that $\{y^\varepsilon\}$ is bounded in $L^\infty(0,T;H^2(\Omega))$ by the method of the translations parallel to the boundary, due to Agmon, Douglis, Nirenberg [5].

For the sake of simplicity, we assume that

$$\Omega = \{x = (x', x_N) \in R^N;\ x_N > 0\}.$$

We denote

$$\delta y(x) = (y(x + he_j) - y(x))/h, \delta^* y(x) = (y(x - he_j) - y(x))/h, 1 \leq j \leq N-1,$$

where $\{e_j\}$, $j = 1,N$, is a basis in R^N, $h > 0$.

Multiply (3.49) by $\delta^*\delta(y_t^\varepsilon)$ and use the Green identity:

$$1/2\ d/dt \int_\Omega |y_t^\varepsilon|^2 dx + 1/2\ d/dt \int_\Omega |\text{grad}\,\delta y^\varepsilon|^2 dx \leq \int_\Omega Bu_\varepsilon\, \delta^*\delta(y_t^\varepsilon) dx.$$

We integrate over $[0,t]$ and make $h \to 0$:

$$1/2\int_\Omega |\partial y_t^\varepsilon/\partial x_j|^2 dx - 1/2\int_\Omega |\partial v_o/\partial x_j|^2 dx + 1/2\int_\Omega \sum_{i=1}^N |\partial^2 y^\varepsilon/\partial x_i \partial x_j|^2 -$$

$$- 1/2 \sum_{i=1}^N \int_\Omega |\partial^2 y_o/\partial x_i \partial x_j|^2 dx \leq \int_0^t \int_\Omega \partial(Bu_\varepsilon)/\partial x_j\, \partial y_t^\varepsilon/\partial x_j\, dx d\tau$$

Let j take all the values from 1 to N-1. Then we see that all the second order derivatives of y^ε, except $\partial^2 y^\varepsilon/\partial x_N^2$, are bounded in $L^\infty(0,T;L^2(\Omega))$. However, since $\{\Delta y^\varepsilon\}$ is bounded in $L^\infty(0,T;L^2(\Omega))$, we get that $\{\partial^2 y^\varepsilon/\partial x_N^2\}$ is bounded in $L^\infty(0,T;L^2(\Omega))$ too and, consequently, $\{y^\varepsilon\}$ is bounded in $L^\infty(0,T;H^2(\Omega))$.

We can use the Aubin compactness result to pass to the limit. $H^2(\Omega)$ is compactly embedded in $H^{7/4}(\Omega)$ and $\{y^\varepsilon\}$, $\{y_t^\varepsilon\}$ are bounded in $L^\infty(0,T;H^2(\Omega))$, $L^\infty(0,T;H^1(\Omega))$. Therefore, on a subsequence, we have $y^\varepsilon \to y$ strongly in $L^2(0,T;H^{7/4}(\Omega))$.

The trace theorem (Ch.I, §1) gives $\partial y^\varepsilon/\partial n \to \partial y/\partial n$ strongly in $L^2(\Sigma)$.

Similarly, by $\{y_t^\varepsilon\}$, $\{y_{tt}^\varepsilon\}$ bounded in $L^\infty(0,T;H^1(\Omega))$, $L^2(0,T;L^2(\Omega))$, we deduce $y_t^\varepsilon|_\Sigma \to y_t|_\Sigma$, strongly in $L^2(\Sigma)$.

Since the maximal monotone operators are closed, we obtain $-\partial y/\partial n \in \beta(y_t)$ a.e. Σ and we can pass to the limit. Obviously $y = (S \circ \theta)(Bu)$ and ,by the uniqueness of the solution, the convergence is true on the initial sequence.

Lemma 3.10. The operator $S \circ \theta_\varepsilon : L^2(0,T;H_0^1(\Omega)) \to L^2(0,T;L^2(\Omega))$ is Gateaux differentiable and $r = \nabla(S \circ \theta_\varepsilon)(f)g$ is the unique generalized solution of the equation

$$\begin{aligned} &(3.58) \quad && r_{tt} - \Delta r = g && \text{in } Q, \\ &(3.59) \quad && r(0,x) = r_t(0,x) = 0 && \text{in } \Omega, \\ &(3.60) \quad && -\partial r/\partial n = \dot\beta^\varepsilon(y_t)r_t && \text{in } \Sigma. \end{aligned}$$

Moreover $r \in L^\infty(0,T;H^1(\Omega))$, $r_t \in L^\infty(0,T;L^2(\Omega))$, $r_t|_\Sigma \in L^2(\Sigma)$.

Proof

We denote $y^\lambda = (S \circ \theta_\varepsilon)(f + \lambda g)$, $y = (S \circ \theta_\varepsilon)(f)$ with $f, g \in L^2(0,T;H^1_o(\Omega))$. We have

$$\begin{aligned} &y^\lambda_{tt} - \Delta y^\lambda = f + \lambda g && \text{in } Q, \\ &y_{tt} - \Delta y = f && \text{in } Q, \end{aligned}$$

with the same boundary and initial conditions.

Subtract the equations and multiply by $y^\lambda_t - y_t$:

$$(3.61) \quad 1/2\, d/dt|y^\lambda_t - y_t|^2_{L^2(\Omega)} + 1/2\, |d/dt \operatorname{grad}(y^\lambda - y)|^2_{L^2(\Omega)} + \int_\Gamma (\beta^\varepsilon(y^\lambda_t) - \beta^\varepsilon(y_t))(y^\lambda_t - y_t)d\sigma \le \lambda\int_\Omega g(y^\lambda_t - y_t)dx.$$

It yields that $\{(y^\lambda - y)/\lambda\}$, $\{(y^\lambda_t - y_t)/\lambda\}$ are bounded in $L^\infty(0,T;H^1(\Omega))$, $L^\infty(0,T;L^2(\Omega))$. On a convenient subsequence, we get $(y^\lambda - y)/\lambda \to r$ strongly in $L^2(Q)$, $y^\lambda \to y$ strongly in $L^2(0,T;H^1(\Omega))$, $y^\lambda_t \to y_t$ strongly in $L^\infty(0,T;L^2(\Omega))$.

By assumption (3.53) and by (3.61) we see that $(y^\lambda_t - y_t)/\lambda$ is bounded in $L^2(\Sigma)$. As

$$(y^\lambda - y)/\lambda|_\Sigma \to r|_\Sigma \qquad \text{weakly in } L^\infty(0,T;H^{1/2}(\Gamma)),$$

a simple distribution argument gives that

$$(y^\lambda_t - y_t)/\lambda|_\Sigma \to r_t|_\Sigma \qquad \text{weakly in } L^2(\Sigma),$$

due to the above boundedness. Now, it is quite standard to show that

$$(\beta^\varepsilon(y^\lambda_t) - \beta^\varepsilon(y_t))/\lambda \to \dot\beta^\varepsilon(y_t)r_t$$

weakly in $L^2(\Sigma)$. We can pass to the limit and prove that r satisfies (3.58)-(3.60) as a generalized solution (Mikhailov [77])

$$\int_Q(\operatorname{grad} r \cdot \operatorname{grad} v - r_t \cdot v_t)dxdt + \int_\Sigma \dot\beta^\varepsilon(y_t)r_t v d\sigma dt = \int_Q gvdxdt$$

for any $v \in L^\infty(0,T;H^1(\Omega)) \cap W^{1,\infty}(0,T;L^2(\Omega))$, $v(x,T) = 0$ a.e. Ω. Obviously (3.58) is satisfied in the sense of distributions. To check the uniqueness of the generalized solution, we take $g = 0$, we integrate over $[t,t+h]\subset[0,T]$ and we multiply by $(r(t+h)-r(t))/h^2$:

$$0=\int_0^t\int_\Omega (r_t(t+h)-r_t(t))/h\,.\,(r(t+h)-r(t))/h\,dxdt-\int_0^t\int_\Omega \Delta 1/h\int_t^{t+h} r\,.\,(r(t+h)-r(t))/h\;dxdt.$$

We integrate by parts

$$1/2|(r(t+h)-r(t))/h|_H^2-1/2|(r(h)-r(0))/h|_H^2+\int_0^t\int_\Omega \text{grad}\,1/h\int_t^{t+h} r\,.\,\text{grad}(r(t+h)-r(t))/h\,-$$

$$-\int_0^t\int_\Gamma \partial/\partial n\,1/h\int_t^{t+h} r(r(t+h)-r(t))/h=0.$$

Therefore

$$1/2|(r(t+h)-r(t))/h|_H^2-1/2|(r(h)-r(0))/h|_H^2+1/2|\text{grad}\,1/h\int_t^{t+h} r|_H^2\,-$$

$$-1/2|\text{grad}\,1/h\int_0^h r|_H^2-\int_0^t\int_\Gamma \partial/\partial n\,1/h\int_t^{t+h} r\,.\,(r(t+h)-r(t))/h=0.$$

We make $h\to 0$ and we get

$$1/2|r_t(t)|_H^2+1/2|\text{grad}\,r(t)|_H^2+\int_\Sigma \dot\beta^\varepsilon(y_t)r_t^2\,d\sigma dt=0,$$

that is $r=0$. a.e. Q.

Let us now analyse the adjoint operator $\nabla(S\circ\theta_\varepsilon)(f)^*: L^2(Q)\to L^2(0,T;H^{-1}(\Omega))$. We denote $p=\nabla(S\circ\theta_\varepsilon)(f)^*q$, where $f\in L^2(0,T;H_0^1(\Omega))$, $q\in L^2(Q)$, $p\in L^2(0,T;H^{-1}(\Omega))$ and we define the adjoint by $p=-m_t$ where, formally, m satisfies

(3.62) $\quad m_{tt}-\Delta m=-\int_t^T q \quad$ in Q,

(3.63) $\quad m(T,x)=m_t(T,x)=0 \quad$ in Ω,

(3.64) $\quad \partial m/\partial n=\dot\beta^\varepsilon(y_t)m_t \quad$ in Σ,

with $y=(S\circ\theta_\varepsilon)(f)$.

Proposition 3.11. Under assumption (3.53), the problem (3.62) - (3.64) has a unique generalized solution $m\in L^\infty(0,T;H^1(\Omega))$, $m_t\in L^\infty(0,T;L^2(\Omega))$.

Proof

Let us denote by $w=-\int_t^T q\in W^{1,2}(0,T;L^2(\Omega))$ and by y^n a regularization from $C^\infty(\bar Q)$ of $y=(S\circ\theta_\varepsilon)(f)$. We consider the approximate equation

(3.65) $\quad m_{tt}^\lambda-\Delta m^\lambda=w \quad$ in Q,

(3.66) $\quad m^\lambda(T,x)=m_t^\lambda(T,x)=0 \quad$ in Ω,

(3.67) $\quad \partial m^\lambda/\partial n=(\beta^\varepsilon(y_t^n+\lambda m_t^\lambda)-\beta^\varepsilon(y_t^n))/\lambda \quad$ in Σ,

for $\lambda>0$.

Formally, we have

$$(y^n + \lambda m^\lambda)_{tt} - \Delta (y^n + \lambda m^\lambda) = f^n + \lambda w$$
$$(y^n + \lambda m^\lambda)(T,x) = y^n(T,x), \quad (y^n + \lambda m^\lambda)_t(T,x) = y^n_t(T,x)$$
$$\partial/\partial n(y^n + \lambda m^\lambda) = \beta^\varepsilon(y^n_t + \lambda m^\lambda_t) - \beta^\varepsilon(y^n_t) + \partial y^n/\partial n,$$

where $f^n = y^n_{tt} - \Delta y^n$.

We write it as a system for the pair $[v^\lambda, p^\lambda]$:

$$v^\lambda_t = p^\lambda,$$
$$p^\lambda_t - \Delta v^\lambda = f^n + \lambda w,$$
$$v^\lambda(T,x) = y^n(T,x), \ p^\lambda(T,x) = y^n_t(T,x),$$
$$\partial v^\lambda/\partial n = \beta^\varepsilon(p^\lambda) - \beta^\varepsilon(y^n_t) + \partial y^n/\partial n.$$

Let $z^\lambda = v^\lambda - h$ where h is choosen such that $h_t, \Delta h \in W^{1,2}(0,T;H^1(\Omega))$ and

$$\partial h/\partial n = -\beta^\varepsilon(y^n_t) \quad + \partial y^n/\partial n \qquad \text{on} \ \Sigma,$$
$$h(T,x) = -a(x) \qquad \text{in} \ \Omega,$$

with $a \in H^2(\Omega)$, $\partial a/\partial n = \beta^\varepsilon(y^n_t(T,x)) - \partial y^n/\partial n(T,x)$ on Γ.

To show the existence of such a h we may take the parabolic problem

$$u_t + \Delta u = 0 \qquad \text{in} \ Q,$$
$$\partial u/\partial n = \dot\beta^\varepsilon(y^n_t) y^n_{tt} - \partial y^n_t/\partial n \qquad \text{in} \ \Sigma,$$
$$u(T,x) = q \qquad \text{in} \ \Omega,$$

where $q \in H^2(\Omega)$ is choosen such that the compatibility conditions

$$\partial q/\partial n = (\dot\beta^\varepsilon(y^n_t) y^n_{tt} - \partial y^n_t/\partial n)(T,x), \quad x \in \Gamma$$

are satisfied. Therefore, a wellknown result (see for instance Barbu [13], p.145) shows that $u \in W^{1,2}(0,T;H^1(\Omega))$ and a distribution argument implies that $\Delta u \in L^2(0,T;H^1(\Omega))$.

Consequently, $h = \int_t^T u - a$ satisfies all the required properties.

The pair $[z^\lambda, p^\lambda]$ satisfies, formally, the system:

$$z^\lambda_t - p^\lambda = -h_t,$$
$$p^\lambda_t - \Delta z^\lambda = f^n + \lambda w + \Delta h,$$
$$\partial z^\lambda/\partial n = \beta^\varepsilon(p^\lambda),$$
$$z^\lambda(T,x) = y^n(T,x) + a, \quad p^\lambda(T,x) = y^n_t(T,x).$$

We check the compatibility conditions

$$\partial y^n(T,x)/\partial n + \partial a/\partial n = \beta^\varepsilon(y_t^n(T,x)) = \beta^\varepsilon(p^\lambda(T,x))$$

and, according to Brezis [20], p.147, the above system has a unique strong solution $[z^\lambda, p^\lambda]$ since the right-hand side is differentiable with respect to t.

Going back with the argument, we obtain a strong solution m^λ of (3.65)-(3.67).

We multiply (3.65) by m_t^λ and integrate by parts:

$$\text{(3.68)} \qquad -1/2\,|m_t^\lambda(t)|_H^2 - 1/2\,|\text{grad}\, m^\lambda(t)|_H^2 - 1/\lambda \int_t^T [\beta^\varepsilon(y_t^n + \lambda m_t^\lambda) - \beta^\varepsilon(y_t^n)] \,.\, m_t^\lambda = \int_t^T w m_t^\lambda.$$

By (3.53) we get $\{m_t^\lambda\}$ bounded in $L^\infty(0,T;L^2(\Omega))$, $\{m^\lambda\}$ bounded in $L^\infty(0,T;H^1(\Omega))$ and $\{m_t^\lambda|_\Sigma\}$ bounded in $L^2(\Sigma)$ with respect to $\lambda > 0$. We have, for $\lambda \to 0$, the convergences

$$m^\lambda \to m^n \qquad \text{weakly* in } L^\infty(0,T;H^1(\Omega)),$$
$$m_t^\lambda \to m_t^n \qquad \text{weakly* in } L^\infty(0,T;L^2(\Omega)),$$
$$(\beta^\varepsilon(y_t^n + \lambda m_t^\lambda) - \beta^\varepsilon(y_t^n))/\lambda \to \dot\beta^\varepsilon(y_t^n) m_t^n \qquad \text{weakly in } L^2(\Sigma).$$

We infer that m^n is a generalized solution of (3.62) - (3.64) with y replaced by y^n. Moreover, by the weak lower semicontinuity of the norm, the estimates given by (3.68) are also valid for m^n. It is easy to make $n \to \infty$ and to obtain a generalized solution of (3.62) - (3.64). The uniqueness may be proved as in Lemma 3.10.

Remark 3.12. The above approach may be used in various situations. For instance, in Proposition 3.6 or Lemma 3.7, we could argue similarly.

Lemma 3.13. Let $y^\varepsilon = (S \circ \theta_\varepsilon)(f)$ and $y = (S \circ \theta)(f)$. Then:

$$\text{(3.69)} \qquad |y_t^\varepsilon - y_t|^2_{L^\infty(0,T;L^2(\Omega))} + |y^\varepsilon - y|^2_{L^\infty(0,T;H^1(\Omega))} \le C\,.\,\varepsilon,$$

where C is a constant independent of $\varepsilon > 0$ and $f \in L^2(0,T;H_0^1(\Omega))$.

Proof

We subtract the equations corresponding to y^ε, y^λ and multiply by $y_t^\lambda - y_t^\varepsilon$:

$$1/2\, d/dt\, |y_t^\varepsilon - y_t^\lambda|^2_{L^2(\Omega)} + 1/2\, d/dt\, |\text{grad}(y^\varepsilon - y^\lambda)|^2_{L^2(\Omega)} + \int_{\partial\Omega} (\beta^\varepsilon(y_t^\varepsilon) - \beta^\lambda(y_t^\lambda)) \,.\, (y_t^\varepsilon - y_t^\lambda) d\sigma = 0.$$

Since $\{\beta^\varepsilon(y_t^\varepsilon)\}$ and $\{\alpha^\varepsilon(y_t^\varepsilon)\}$ are bounded in $L^2(\Sigma)$ from (3.54) a calculus similar to Lemma 3.5 gives (3.69).

Proposition 3.14. Under condition (3.53) the problem (3.55) has at least one optimal pair $[y^\varepsilon, u_\varepsilon]$ in $W^{2,2}(0,T;L^2(\Omega)) \times L^2(0,T;U)$ and there is $m^\varepsilon \in L^\infty(0,T;H^1(\Omega))$ such that

$$y_{tt}^\varepsilon - \Delta y^\varepsilon = B u_\varepsilon \qquad \text{in } Q,$$
$$\text{(3.70)} \qquad m_{tt}^\varepsilon - \Delta m^\varepsilon = -\int_t^T q_\varepsilon \qquad \text{in } Q,$$
$$y^\varepsilon(0,x) = y_0(x), \quad y_t^\varepsilon(0,x) = v_0(x) \qquad \text{in } \Omega,$$
$$\text{(3.71)} \qquad m^\varepsilon(T,x) = m_t^\varepsilon(T,x) = 0 \qquad \text{in } \Omega,$$

$$-\partial y^\varepsilon/\partial n = \beta^\varepsilon(y_t^\varepsilon) \qquad \text{in } \Sigma,$$

(3.72) $$\partial m^\varepsilon/\partial n = \dot\beta^\varepsilon(y_t^\varepsilon)m_t^\varepsilon \qquad \text{in } \Sigma,$$

$$[q_\varepsilon(t), -B^*m_t^\varepsilon(t) + u_\varepsilon(t) - u^*(t)] = \partial L^\varepsilon(y^\varepsilon(t), u_\varepsilon(t)) \qquad \text{in } [0,T].$$

Moreover $y^\varepsilon \to y^*$ strongly in $C(0,T;H^1(\Omega))$, $y_t^\varepsilon \to y_t^*$ strongly in $C(0,T;L^2(\Omega))$, $u_\varepsilon \to u^*$ strongly in $L^2(0,T;U)$, $p^\varepsilon = -m_t^\varepsilon \to p^*$ weakly* in $L^\infty(0,T;L^2(\Omega))$, $q_\varepsilon \to q^*$ weakly in $L^1(0.T;L^2(\Omega))$ and

(3.73) $$[q^*(t), B^*p(t)] \in \partial L(y^*(t),u^*(t)) \qquad \text{a.e. } [0,T].$$

This is a direct application of Thm.1.6 and of Lemmas 3.9 - 3.13.

To pass to the limit in (3.70) - (3.72) we impose again conditions (2.5), (3.27).

Theorem 3.15. Let $[y^*,u^*] \in W^{2,2}(0,T;L^2(\Omega)) \times L^2(0,T;U)$ be an optimal pair for the problem (3.48) - (3.51), where β is a maximal monotone graph in $R \times R$, satisfying (3.53), (2.5), (3.27). Then, there exist $m^* \in L^\infty(0,T;H^1(\Omega)) \cap W^{1,\infty}(0,T;L^2(\Omega))$, $p^* = -m_t^*$ and $q^* \in L^2(Q)$ such that

$$m_{tt}^* - \Delta m^* = -\int_t^T q^* \qquad \text{in } Q,$$

$$m^*(T,x) = m_t^*(T,x) = 0 \qquad \text{in } \Omega,$$

$$\partial m^*/\partial n \in D\beta(y_t^*)m_t^* \qquad \text{in } \Sigma,$$

in a weak sense. The functions y^*,u^*,m^*,q^* are limits of the mappings $y^\varepsilon, u_\varepsilon, m^\varepsilon, q_\varepsilon$ given by the Proposition 3.14.

Proof

We remark that the estimates given by (3.68) are also independent of $\varepsilon > 0$, therefore $\{m^\varepsilon\}$ is bounded in $L^\infty(0,T;H^1(\Omega))$, $\{m_t^\varepsilon\}$ is bounded in $L^\infty(0,T;L^2(\Omega))$, $\{m_t^\varepsilon|_\Sigma\}$ is bounded in $L^2(\Sigma)$. It is easy to infer that $\{\dot\beta^\varepsilon(y_t^\varepsilon)|m_t^\varepsilon|^2\}$ is bounded in $L^1(\Sigma)$ too. Since $\{\beta^\varepsilon(y_t)\}$ is bounded in $L^2(\Sigma)$, hypothesis (2.5) implies that $\{\dot\beta^\varepsilon(y_t^\varepsilon)\}$ is also bounded in $L^2(\Sigma)$. Next, the inequality of Young shows as in the proof of Thm. 3.8, that $\{\dot\beta^\varepsilon(y_t^\varepsilon)m_t^\varepsilon\}$ is bounded in $L^{4/3}(\Sigma)$ and $\dot\beta^\varepsilon(y_t^\varepsilon)m_t^\varepsilon \to z$ weakly in $L^{4/3}(\Sigma)$. By means of a saddle function, as in the preceding paragraph, one can see that

$$z(t,x) \in D\beta(y_t^*(t,x))m_t^*(t,x), \quad \text{a.e. } \Sigma$$

and this ends the proof.

Remark 3.16. A similar result may be established when has good differentiability properties. In the recent work [30] of Brokate and Sprekels, optimality conditions for a control problem governed by a second order nonlinear evolution equation (with a fourth order elliptic operator) coupled with a nonlinear parabolic equation are obtained under differentiability assumptions. This is related to a model describing thermomechanical solid-solid phase transitions in a piece of metallic alloy.

Remark 3.17. Returning to the abstract problem (3.1) - (3.3), if L is a quadratic cost functional and no constraints are imposed, we get $u^* = B^*p^*$ both in § 3.1 and in § 3.2, by (3.23), (3.73). Then all the optimal controls of the problem (3.1) - (3.3) satisfy $u^* \in L^\infty(0,T;U)$.

4. Quasilinear problems

In this section we extend the previous results to unconstrained control problems governed by general nonlinear state equations:

(4.1) Minimize $\{G(y) + F(u)\}$,

(4.2) $y_{tt} - \Delta y + f(y, y_t, \text{grad } y) = u$ in Q,

(4.3) $y(0,x) = y_0(x),\ y_t(0,x) = v_0(x)$ in Ω,

(4.4) $y(t,x) = 0$ in Σ.

One may consider more general differential operators or boundary conditions. Parabolic problems with nonlinear terms of the form $g(y, \text{grad } y)$ may be also included in our discussion. However, for the sake of brevity, we limit ourselves to the examination of the problem (4.1) - (4.4).

We assume that $F,G : L^2(Q) \to R$ are convex, continuous and $y_0 \in H_0^1(\Omega)$, $v_0 \in L^2(\Omega)$, $u \in L^2(Q)$. The mapping $f : R^{N+2} \to R$ is locally Lipschitzian and satisfies the condition: for any $M > 0$, there is $C > 0$ such that

(4.5) $$|f(y) - f(z)| \le C(1 + |f(y)|)|y - z|,\quad |y - z| \le M,$$

where $|\cdot|$ is the norm in R^{N+2} or the modulus, as appropriate. This is a slight strengthening of the condition (2.5). It also admits a large class of examples including polynomials or exponentials.

In the absence of the monotonicity hypothesis, which played an essential role in the previous sections, the system (4.2) - (4.4) may be not well posed: this is a singular or an unstable state system. For instance, we may have only local existence or the solution may be not unique.

The problem (4.1) - (4.4) enters the setting of singular control problems. The first results for this type of problems are due to J.L.Lions [72], J.F.Bonnans [22].In our presentation, we consider general state equations and cost functionals, under weak differentiability conditions. The method we use combines the adapted penalization idea, V.Barbu [13] with the idea to penalize the nonlinear term in the cost functional, J.L.Lions [72], and consists in an adapted penalization of the nonlinear term in the approximate problem.

To clarify this discussion, we start with an example of singular system.

Let $f : R \to R$, $f(y) = -y^3$ (it is not monotone) and $u = 0$ a.e. Q. We assume that the initial conditions (4.3) satisfy $(y_0, v_0)_{L^2(\Omega)} > 0$ and

(4.6) $$E^0 = 1/2\,|\text{grad } y_0|^2_{L^2(\Omega)} + 1/2\,|v_0|^2_{L^2(\Omega)} - 1/4\int_\Omega |y_0|^4 dx \le 0.$$

<u>Proposition 4.1</u> (Lions [72], Ch.2,§1.4). <u>Under the above hypotheses, the equation</u> (4.2) - (4.4) <u>has no solution</u> $y \in L^6(Q)$.

<u>Proof</u>

We denote $E(t) = |y(t)|^2_{L^2(\Omega)}$ and we compute formally:

$$E'(t) = 2(y(t), y_t(t))_{L^2(\Omega)},$$
$$E''(t) = 2(y(t), y_{tt}(t))_{L^2(\Omega)} + 2|y_t(t)|^2_{L^2(\Omega)} =$$
$$= 2(y(t), \Delta y(t) + y(t)^3)_{L^2(\Omega)} + 2|y_t(t)|^2_{L^2(\Omega)} =$$
$$2[|y_t(t)|^2_{L^2(\Omega)} - |\mathrm{grad}\, y(t)|^2_{L^2(\Omega)} + \int_\Omega |y(t)|^4 dx].$$

We multiply (4.2) by y_t and obtain

$$1/2d/dt[|y_t(t)|^2_{L^2(\Omega)} + |\mathrm{grad}\, y(t)|^2_{L^2(\Omega)} - 1/2\int_\Omega |y(t)|^4 dx] = 0.$$

Therefore, the energy of the system is constant with respect to the time and we get

$$\int_\Omega |y(t)|^4 dx = 2[|y_t(t)|^2_{L^2(\Omega)} + |\mathrm{grad}\, y(t)|^2_{L^2(\Omega)}] - 4E^0.$$

It yields that

$$E''(t) = 2[|y_t(t)|^2_{L^2(\Omega)} - |\mathrm{grad}\, y(t)|^2_{L^2(\Omega)} + 2|y_t(t)|^2_{L^2(\Omega)} +$$
$$+ 2|\mathrm{grad}\, y(t)|^2_{L^2(\Omega)} - 4E^0] \geq 6|y_t(t)|^2_{L^2(\Omega)}.$$

Consequently,

$$E(t)\,.\,E''(t) - 3/2E'(t)^2 \geq 6[|y(t)|^2_{L^2(\Omega)}\cdot|y_t(t)|^2_{L^2(\Omega)} - (y(t),y_t(t))^2_{L^2(\Omega)}] \geq 0.$$

This is equivalent with $[E^{-1/2}(t)]'' \leq 0$ and, integrating over $[0,t]$, we infer

$$E^{-1/2}(t) \leq t(E^{-1/2})'(0) + E^{-1/2}(0) = E^{-3/2}(0)(E(0) - 1/2tE'(0)).$$

Since $E(t)$ is positive, we obtain

(4.7) $$E^{1/2}(t) \geq E^{3/2}(0)(E(0) - 1/2tE'(0))^{-1}.$$

The condition $(y_0, v_0)_{L^2(\Omega)} > 0$ gives $E'(0) > 0$ and we define $t_0 > 0$ such that

(4.8) $$E(0) = t_0/2E'(0).$$

Then (4.7), (4.8) imply

$$\int_0^{t_0} |y|_{L^2(\Omega)} dt = + \quad .$$

If (4.2) - (4.4) has a solution in $L^6(Q)$, then $y^3 \in L^2(Q)$ and we have $y \in L^\infty(0,T;H^1_0(\Omega)) \cap W^{1,\infty}(0,T;L^2(\Omega))$. The above operations are justified and if T is sufficiently big with respect to t_0 we arrive at a contradiction.

Remark 4.2. It is possible to indicate situations when (4.2) - (4.4) has a global solution. For instance, again for $f(y) = -y^3$, but for zero initial data, if $u = 0$, then we have $y = 0$.

Generally, for fixed initial data, we call the pair $[y,u] \in L^2(Q) \times L^2(Q)$ to be admissible for the problem (4.1) - (4.4) if y is the solution of (4.2), corresponding to u, in a convenient sense.

For simplicity, we ask that $y \in H^1(Q)$, $f(y, y_t, \text{grad } y) \in L^2(Q)$ and y is the generalized solution of (4.2) corresponding to u (see § 3.2).

Therefore, the singular control problem (4.1) - (4.4) consists in finding the optimal pair $[y^*, u^*]$ which minimizes the cost functional (4.1) over the set of all the admissible pairs.

In certain conditions, it is possible to show that the set of admissible pairs is "rich" and, consequently, the singular control problem is not trivial.

Proposition 4.3 (Bonnans [22], Ch.3)

Let $y_0 = v_0 = 0$, $f = -y^3$ and $\Omega \subset R^3$ be a bounded domain. Then, the set of admissible controls

$$O = \left\{ u \in L^2(Q);\ (4.2) - (4.4) \text{ has a generalized solution } y \in H^1(Q), \text{ and } f(y) \in L^2(Q) \right\}$$

is an open, nonvoid, connected subset of $L^2(Q)$.

Proof

O is nonvoid by Remark 4.2. Let

$$W_T = \left\{ y \in H^1(Q);\ y'' - \Delta y \in L^2(Q);\ y(0,x) = y_t(0,x) = 0 \quad \text{a.e.} \Omega, y(t,x) = 0 \quad \text{on } \Sigma \right\}.$$

By the regularity results for hyperbolic equations, we obtain $y \in L^\infty(0,T;H_0^1(\Omega))$, $y_t \in L^\infty(0,T;L^2(\Omega))$.

It is known that, for $\Omega \subset R^3$, $H_0^1(\Omega) \subset L^6(\Omega)$, therefore $f(y) \in L^2(Q)$, for $y \in W_T$. We conclude that W_T is the space of admissible states and the mapping from W_T to $L^2(Q)$, given by

$$y \to y_{tt} - \Delta y - y^3$$

is continuous and O is its image. It yields that O is connected in $L^2(Q)$.

For $u \in O$, the solution of (4.2) - (4.4) is unique. To see this we proceed by contradiction. Let y_1, y_2 be two generalized solutions of (4.2) - (4.4) from W_T. We subtract the equations and multiply by $(y_1 - y_2)_t$

$$1/2\, d/dt |(y_1 - y_2)_t(t)|^2_{L^2(\Omega)} + 1/2\, d/dt |\text{grad}(y_1 - y_2)(t)|^2_{L^2(\Omega)} = \int_\Omega (y_1^3(t) - \tag{4.9}$$

$$- y_2^3(t))(y_1 - y_2)_t(t)dx \leq | y_1^2(t) + y_1(t)y_2(t) + y_2^2(t)|_{L^3(\Omega)} \, | y_1(t) - y_2(t)|_{L^6(\Omega)}$$

$$|(y_1 - y_2)_t|_{L^2(\Omega)} \leq C\left\{ |y_1(t) - y_2(t)|^2_{H_0^1(\Omega)} + |(y_1 - y_2)_t(t)|^2_{L^2(\Omega)} \right\},$$

since $y_i \in L^\infty(0,T;H_0^1(\Omega))$ and $\Omega \subset R^3$.

The Gronwall inequality implies $y_1 = y_2$.

The mapping $A : W_T \times L^2(Q) \to L^2(Q)$, given by

$$A(y,u) = y_{tt} - \Delta y - y^3 - u$$

is obviously of class C^1 and $[\partial A/\partial y](y;u) : z \to z_{tt} - \Delta z - 3y^2 z$ is an isomorphism from W_T to $L^2(Q)$, according to Lemma 4.4 below. The implicit mapping theorem shows that the function $u \to y(u)$ is regular and O is open.

Lemma 4.4. Let $q \in L^\infty(0,T;L^3(\Omega))$ and $v \in L^2(Q)$. The equation

(4.10) $\quad y_{tt} - \Delta y + qy = v \quad$ in Q,

(4.11) $\quad y(0,x) = y_t(0,x) = 0 \quad$ in Ω,

(4.12) $\quad y(t,x) = 0 \quad$ in Σ,

has a unique solution in W_T.

Proof

The existence and the regularity of the solution may be obtained by approximating q with an $L^\infty(Q)$ function and by estimates of the type (4.9). The uniqueness may be proved by contradiction, again by (4.9). See also Lions [72], Ch.2, Theorem 2.2, for a more general result.

Let us return to the problem (4.1) - (4.4). By the above discussion we see the necessity of an admissibility condition: the problem (4.1) - (4.4) has at least one admissible pair [y,u], in the sense of Remark 4.2.

An important feature of singular control problems is that, although the state system is not well posed, the optimization problem is well posed. In the problem (4.1) - (4.4), if G satisfies a sufficiently strong coercivity condition, under the admissibility hypothesis, we may show the existence of optimal pairs $[y^*,u^*]$. For instance, in the model situation $f(y)=-y^3$, we take

$$G(y) = \int_Q |y - y_d|^6 dx,$$

$y_d \in L^6(Q)$. Then we obtain the boundedness of a minimizing sequence $[y_n,u_n]$ in $L^6(Q) \times L^2(Q)$ and we may easily prove the existence of at least one optimal pair.

Now, we turn to the main topic of this section, the question of necessary optimality conditions for the problem (4.1) - (4.4).We approximate (4.1) - (4.4) by the optimization problem

(4.13) Minimize

$$G_\varepsilon(y) + F(u) + 1/2\,|f^\varepsilon(y,y_t,\text{grad } y) - f(y^*,y_t^*,\text{grad } y^*)|^2_{L^2(Q)} +$$
$$+ 1/2\,|u-u^*|^2_{L^2(Q)} + 1/2\,|y-y^*|^2_{H^2(Q)} + 1/2\varepsilon\,|y_{tt} - \Delta y + f^\varepsilon(y,y_t,\text{grad } y) - u|^2_{L^2(Q)}.$$

Here G_ε and f^ε are regularizations of the mappings G,f, given by

(4.14) $$G_\varepsilon(y) = \inf\left\{ |z-y|^2_{L^2(Q)}/2\varepsilon + G(z);\ z \in L^2(Q)\right\},$$

(4.15) $$f^{\varepsilon}(z) = \int_{R^{N+2}} f(z - \varepsilon\tau)\rho(\tau)d\tau, \quad z \in R^{N+2},$$

$$\rho \in C_0^{\infty}(R^{N+2}),\ \operatorname{supp}\rho \subset S(0,1),\ \int_{R^{N+2}} \rho(\tau)d\tau = 1,\ \rho(\tau) \geq 0,\ \rho(-\tau) = \rho(\tau).$$

Remark 4.5. The idea to penalize $y - y^*$ in higher order Sobolev spaces was previously used by V.Barbu [16] while the adapted penalization of the nonlinear term appears in [133] and it gives the possibility to consider cost functionals independent of the structure of the state equation.

The pair $[\tilde{y}, \tilde{u}]$ is called ε- admissible (admissible for the problem (4.13)) if $\tilde{u} \in L^2(Q)$, $\tilde{y} - y^* \in H^2(Q)$, $f^{\varepsilon}(\tilde{y}, \tilde{y}_t, \operatorname{grad} \tilde{y}) \in L^2(Q)$, $\tilde{y}(0,x) = y_0(x)$, $\tilde{y}_t(0,x) = v_0(x)$ in Ω and $\tilde{y}(t,x) = 0$ in Σ.

Proposition 4.6. Let $[\bar{y},\bar{u}]$ be ε - admissible for (4.1) - (4.4). Then $[\bar{y} + z, v]$ is ε-admissible, for any $v \in L^2(Q)$, $z \in C^2(\bar{Q})$, $z(0,x) = z_t(0,x) = 0$ in Ω and $z(t,x) = 0$ in Σ. Moreover, the pairs $[y^* + z, v]$ are ε - admissible.

Proof

We have to show that $f^{\varepsilon}(\bar{y} + z,\ \bar{y}_t + z_t,\ \operatorname{grad}(\bar{y} + z))$ is in $L^2(Q)$. We denote $w = (\bar{y} + z,\ y_t + z_t, \operatorname{grad}(y + z))$ and we have

$$|f^{\varepsilon}(w)| \leq \int_{S(0,1)} |f(w - \varepsilon\tau)| \rho(\tau)d\tau \leq$$
$$\leq |f(\bar{y},\bar{y}_t,\operatorname{grad}\bar{y}) + \int_{S(0,1)} |f(w - \varepsilon\tau) - f(\bar{y},\bar{y}_t,\operatorname{grad}\bar{y})| \rho(\tau)d\tau.$$

By (4.5) we deduce that

$$|f^{\varepsilon}(w)| \leq f(\bar{y},\bar{y}_t,\operatorname{grad}\bar{y}) + C\int_{S(0,1)} (1 + |f(\bar{y},\bar{y}_t,\operatorname{grad}\bar{y})|) . |(z,z_t,\operatorname{grad} z) - \varepsilon\tau| \rho(\tau)d\tau$$

since the term $(z,z_t,\operatorname{grad} z) - \varepsilon\tau$ is uniformly bounded on Q.

Proposition 4.7. The problem (4.13) has at least one optimal pair $[y^{\varepsilon},u_{\varepsilon}]$.

Proof

Let $[y^k,u_k]$ be a minimizing sequence. Then $\{u_k\}$ is bounded in $L^2(Q)$, $\{f^{\varepsilon}(y^k,y_t^k,\operatorname{grad} y^k)\}$ is bounded in $L^2(Q)$, $\{y^k - y^*\}$ is bounded in $H^2(Q)$. On a subsequence, we have

$u_k \to \tilde{u}$	weakly in $L^2(Q)$,
$y^k \to \tilde{y}$	strongly in $H^1(Q)$,
$f^{\varepsilon}(y^k,y_t^k,\operatorname{grad} y^k) \to \tilde{f}$	weakly in $L^2(Q)$.

To identify $\tilde{f}$, take any $\eta > 0$ and $Q_\eta \subset Q$ measurable, $\operatorname{mes}(Q \setminus Q_\eta) < \eta$, such that $(y^k, y_t^k, \operatorname{grad} y^k) \to (\tilde{y},\tilde{y}_t,\operatorname{grad}\tilde{y})$ uniformly on Q_η, by the Egorov theorem. Then (4.5) gives $f^{\varepsilon}(y^k,y_t^k, \operatorname{grad} y^k)$ $f^{\varepsilon}(\tilde{y},\tilde{y}_t,\operatorname{grad}\tilde{y})$ uniformly on Q_η, so $\tilde{f} = f^{\varepsilon}(\tilde{y},\tilde{y}_t,\operatorname{grad}\tilde{y})$ a.e. Q.

Obviously $[\tilde{y},\tilde{u}]$ is ε-admissible and, by the weak lower semicontinuity of $G + F$, it is ε - optimal. We denote it $[y^{\varepsilon},u_{\varepsilon}]$.

Lemma 4.8. For $\varepsilon \to 0$, we have

(4.16) $$u_{\varepsilon} \to u^* \qquad \text{strongly in } L^2(Q),$$

(4.17) $y^\varepsilon - y^* \to 0$ <u>strongly in</u> $H^2(Q)$,

(4.18) $f^\varepsilon(y^\varepsilon, y_t^\varepsilon, \text{grad } y^\varepsilon) \to f(y^*, y_t^*, \text{grad } y^*)$ <u>strongly in</u> $L^2(Q)$.

<u>Proof</u>

Denote by J_ε the functional (4.13).Then

(4.19) $$J_\varepsilon(y^\varepsilon, u_\varepsilon) \leq J_\varepsilon(y^*, u^*) = G_\varepsilon(y^*) + F(u^*) + 1/2(1 + 1/\varepsilon) . | f^\varepsilon(y^*, y_t^*, \text{grad } y^*) - f(y^*, y_t^*, \text{grad } y^*)|^2_{L^2(Q)} \leq G_\varepsilon(y^*) + F(u^*) + C(1 + 1/\varepsilon)\varepsilon^2(1 + | f(y^*, y_t^*, \text{grad } y^*)|^2_{L^2(Q)}),$$

since, by (4.5), we get

(4.20) $$|f^\varepsilon(y) - f(y)| \leq C\varepsilon(1 + |f(y)|), \quad \varepsilon \in [0,1], \ y \in R^{N+2}.$$

We obtain

(4.21) $$\limsup_{\varepsilon \to 0} J_\varepsilon(y^\varepsilon, u_\varepsilon) \leq G(y^*) + F(u^*).$$

Consequently, $\{u_\varepsilon\}$ is bounded in $L^2(Q)$, $\{y^\varepsilon - y^*\}$ is bounded in $H^2(Q)$, $\{f^\varepsilon(y^\varepsilon, y_t^\varepsilon, \text{grad } y^\varepsilon)\}$ is bounded in $L^2(Q)$ and we have the convergence in $L^2(Q)$

(4.22) $$y_{tt}^\varepsilon - \Delta y^\varepsilon + f^\varepsilon(y^\varepsilon, y_t^\varepsilon, \text{grad } y^\varepsilon) - u_\varepsilon \to 0.$$

An argument similar to <u>Proposition 4.7</u> shows that $f^\varepsilon(y^\varepsilon, y_t^\varepsilon, \text{grad } y^\varepsilon) \to f(\hat{y}, \hat{y}_t, \text{grad } \hat{y})$ weakly in $L^2(Q)$, where $[\hat{y}, \hat{u}]$ is the limit of $[y^\varepsilon, u_\varepsilon]$ in the strong-weak topology of $L^2(Q) \times L^2(Q)$.

Then $[\hat{y}, \hat{u}]$ is an admissible pair for (4.1) - (4.4) and we deduce

(4.23) $$\liminf_{\varepsilon \to 0} J_\varepsilon(y^\varepsilon, u_\varepsilon) \geq G(\hat{y}) + F(\hat{u}) + 1/2 |\hat{u} - u^*|^2_{L^2(Q)} + 1/2 |\hat{y} - y^*|^2_{H^2(Q)}.$$

Taking into account (4.21), (4.23) and the optimality of $[y^*, u^*]$, it yields $\hat{y} = y^*$, $\hat{u} = u^*$ and (4.16), (4.17).From (4.19) we remark that

$$G_\varepsilon(y^*) + F(u^*) + C\varepsilon \geq J_\varepsilon(y^\varepsilon, u_\varepsilon) \geq G_\varepsilon(y^\varepsilon) + F(u_\varepsilon) + 1/2 | f^\varepsilon(y^\varepsilon, y_t^\varepsilon, \text{grad } y^\varepsilon) -$$

$$- f(y^*, y_t^*, \text{grad } y^*)|^2_{L^2(Q)} + 1/2 | u_\varepsilon - u^*|^2_{L^2(Q)} + 1/2 | y^\varepsilon - y^*|^2_{H^2(Q)}.$$

Then, as $\varepsilon \to 0$

$$1/2 | f^\varepsilon(y^\varepsilon, y_t^\varepsilon, \text{grad } y^\varepsilon) - f(y^*, y_t^*, \text{grad } y^*)|^2_{L^2(Q)} + 1/2 | u_\varepsilon - u^*|^2_{L^2(Q)} +$$

$$+ 1/2 | y^\varepsilon - y^*|^2_{H^2(Q)} \leq C\varepsilon + G_\varepsilon(y^*) - G_\varepsilon(y^\varepsilon) \to 0$$

<u>Lemma 4.9. Assume that</u> F <u>is continuously Frechet differentiable. There exists</u> $p_\varepsilon \in L^2(Q)$ <u>which satisfy the approximate optimality system</u>

(4.24) $\quad -p_\varepsilon = \nabla F(u_\varepsilon) + u_\varepsilon - u^*$,

(4.25) $\quad \langle p_\varepsilon, \xi_{tt} - \Delta\xi + [\mathrm{grad}\, f^\varepsilon(y^\varepsilon, y^\varepsilon_t, \mathrm{grad}\, y^\varepsilon), (\xi, \xi_t, \mathrm{grad}\,\xi)] \rangle = \langle f^\varepsilon(y^\varepsilon, y^\varepsilon_t, \mathrm{grad}\, y^\varepsilon) -$

$- f(y^*, y^*_t, \mathrm{grad}\, y^*), [\mathrm{grad}\, f^\varepsilon(y^\varepsilon, y^\varepsilon_t, \mathrm{grad}\, y^\varepsilon), (\xi, \xi_t, \mathrm{grad}\,\xi)] \rangle + \langle \partial G_\varepsilon(y^\varepsilon), \xi \rangle +$

$+ \langle y^\varepsilon - y^*, \xi \rangle_{H^2}$,

for all $\xi \in C^2(\bar{Q})$, $\xi(0,x) = \xi_t(0,x) = 0$ in Ω and $\xi(t,x) = 0$ on Σ.

Here we denote by $[.,.]$ the scalar product in R^{N+2}, by $\langle .,. \rangle$ the scalar product in $L^2(Q)$, by $\langle .,. \rangle_{H^2}$ the product in $H^2(Q)$ and by grad f^ε the gradient of f^ε as a function on R^{N+2}.

Proof

We put $p_\varepsilon = -1/\varepsilon\,(y^\varepsilon_{tt} - \Delta y^\varepsilon + f^\varepsilon(y^\varepsilon, y^\varepsilon_t, \mathrm{grad}\, y^\varepsilon) - u_\varepsilon)$. To obtain (4.24) we compute the subdifferential of J_ε with respect to u and we use the minimum property of u_ε. By Proposition 4.6, the pairs $[y^\varepsilon + s\xi, u_\varepsilon]$ are ε- admissible for $s \in R$ and we deduce (4.25) from

$$0 = \lim_{s \to 0} 1/s(J_\varepsilon(y^\varepsilon + s\xi, u_\varepsilon) - J_\varepsilon(y^\varepsilon, u_\varepsilon)).$$

Theorem 4.10. Assume that F is continuously Frechet differentiable. There exist $p^* \in L^2(Q)$ and $\gamma^* \in L^2(Q)^{N+2}$, $\gamma^* \in Df(y^*, y^*_t, \mathrm{grad}\, y^*)$ a.e.Q, such that the following optimality conditions are satisfied

(4.26) $\quad -p^* = \nabla F(u^*)$,

(4.27) $\quad \langle p^*, \xi_{tt} - \Delta\xi + [\gamma^*, (\xi, \xi_t, \mathrm{grad}\,\xi)] \rangle = \langle \partial G(y^*), \xi \rangle$,

for all $\xi \in C^2(\bar{Q})$, $\xi(0,x) = \xi_t(0,x) = 0$ in Ω and $\xi(t,x) = 0$ on Σ.

The notations are the same as in Lemma 4.9.

Proof

From the assumption on F and (4.24), (4.16) we see that $p_\varepsilon \to p^*$ strongly in $L^2(Q)$. The subsequent Lemma 4.11, with technical character, shows that $\{\mathrm{grad}\, f^\varepsilon(y^\varepsilon, y^\varepsilon_t, \mathrm{grad}\, y^\varepsilon)\}$ is bounded in $L^2(Q)^{N+2}$. We use Thm.3.14, Ch.I and we deduce that, on a subsequence

$$\mathrm{grad}\, f^\varepsilon(y^\varepsilon, y^\varepsilon_t, \mathrm{grad}\, y^\varepsilon) \to \gamma^* \in Df(y^*, y^*_t, \mathrm{grad}\, y^*)$$

weakly in $L^2(Q)^{N+2}$. We may pass to the limit in (4.24), (4.25) to obtain (4.26), (4.27).

Lemma 4.11. For ε sufficiently small, we have

(4.28) $\quad \mathrm{grad}|f^\varepsilon(y)| \le C(1 + |f^\varepsilon(y)|), \qquad y \in R^{N+2}$,

with C independent of ε.

Proof

We prove (4.28) for a component i, $1 \le i \le N+2$, of the gradient of f^{ε}. We denote it f_i^{ε}:

(4.29) $$|f_i^{\varepsilon}(y)| \le \lim_{h \to 0} 1/|h| \int_{S(0,1)} |f(y-\varepsilon\tau) - f(y_1 - \varepsilon\tau_1, \dots y_i + h - \varepsilon\tau_i, \dots, y_{N+2} - \varepsilon\tau_{N+2})|$$

$$\rho(\tau)d\tau \le C\int_{S(0,1)} (1 + |f(y-\varepsilon\tau)|)\rho(\tau)d\tau \le C(1 + |f(y)|) +$$

$$+ C\int_{S(0,1)} (1 + |f(y)|)\varepsilon|\tau|\rho(\tau)d\tau \le C(1 + |f(y)|).$$

Here, we denote by C various constants independent of ε and we use several times (4.5).

On the other side, for ε sufficiently small, by (4.20), we have

$$|f(y)| \le |f^{\varepsilon}(y)| + |f^{\varepsilon}(y) - f(y)| \le |f^{\varepsilon}(y)| + 1/2\,(1 + |f(y)|).$$

Combining this with (4.29), we get (4.28).

Remark 4.12. By (4.27) we see that p^* is a weak solution, obtained by the transposition method, of the adjoint system

$$
\begin{array}{ll}
p_t^* = -q^* + \gamma_2^* p^* & \text{in } Q,\\
q_t^* = -\Delta p^* + \gamma_1^* p^* - \operatorname{div}(\gamma_3^* p^*) - \partial G(y^*) & \text{in } Q,\\
p^*(T,x) = q^*(T,x) = 0 & \text{in } \Omega,\\
p^*(t,x) = 0 & \text{in } \Sigma,
\end{array}
$$

where $\gamma_1^* \in L^2(Q)$, $\gamma_2^* \in L^2(Q)$, $\gamma_3^* \in L^2(Q)^N$ are the components of γ^*. The relation (4.26) gives the maximum principle.

Remark 4.13. If f $C^1(R^{N+2})$ it is sufficient to assume that F is continuous and convex on $L^2(Q)$ and Thm.4.10 follows.

From $f \in C^1(R^{N+2})$ we see that grad $f^{\varepsilon}(y^{\varepsilon}, y_t^{\varepsilon}, \text{grad } y^{\varepsilon}) \to$ grad $f(y^*, y_t^*, \text{grad } y^*)$ a.e.Q, by (4.15) and (4.17). Moreover (4.28) and (4.18) show that we may apply the Vitali theorem to pass to the limit in the integral and we obtain

$$\text{grad } f^{\varepsilon}(y^{\varepsilon}, y_t^{\varepsilon}, \text{grad } y^{\varepsilon}) \to \text{grad } f(y^*, y_t^*, \text{grad } y^*)$$

strongly in $L^2(Q)$.

On the other side $\{u_{\varepsilon}\}$ is compact in $L^2(Q)$, $\partial F(.)$ is locally bounded on $L^2(Q)$ and (4.24) implies that $\{p_{\varepsilon}\}$ is bounded in $L^2(Q)$. We can pass to the limit in (4.24), (4.25) and reobtain (4.26), (4.27) with $\gamma^* =$ grad $f(y^*, y_t^*, \text{grad } y^*)$. This is a slight generalization of a result of Komornik and Tiba [123], [124].

5. Other applications

In § 2,3, we have discussed semilinear parabolic or hyperbolic problems, by means of the abstract scheme developped in § 1.The results may be extended to quasilinear problems in the setting of unstable systems control theory, according to § 4.

Now, we present new optimality theorems for nonlinear control problems, which enlarge the domain of application of § 1.

5.1. A nonlinear parabolic problem

Consider the problem

(5.1) Minimize $\int_0^T L(y(t), u(t))dt$,

(5.2) $y_t - \sum_{i=1}^{N} (a_i(y_{x_i}))_{x_i} = Bu$ in Q,

(5.3) $y(0,x) = y_0(x)$ in Ω,

(5.4) $y(t,x) = 0$ in Σ.

We assume that the functions $a_i : R \to R$ satisfy the conditions:

(5.5) $a_i(y)y \geq \omega |y|^p + C, \quad \omega > 0, \quad p \geq 2,$

(5.6) $(a_i(y) - a_i(z))(y - z) \geq \eta |y - z|^2, \quad \eta > 0,$

(5.7) a_i are locally Lipschitzian on R and $\dot{a}_i(y) \leq C(|y|^{p-2} + 1)$a.e.$R$.

From (5.7) it obviously yields $|a(y)| \leq C + C|y|^{p-1}$.Together with (5.5) and with the monotonicity assumption from (5.6), this implies that, if $Bu \in L^{p'}(0,T;W^{-1,p'}(\Omega))$, $1/p + 1/p' = 1$, $y_0 \in L^2(\Omega)$, then the equation (5.2) - (5.3) has a unique solution y in $C(0,T;L^2(\Omega)) \cap L^p(0,T;W_0^{1,p}(\Omega))$ and $y_t \in L^{p'}(0,T;W^{-m,p'}(\Omega))$, Barbu [12], p.145.

Let $a(.,.) : W_0^{1,p}(\Omega) \times W_0^{1,p}(\Omega) \to R$ be the functional

(5.8) $$a(u,v) = \int_\Omega (\sum_{i=1}^{N} a_i(u_{x_i})v_{x_i})dx.$$

It is known that the operator $A : W_0^{1,p}(\Omega) \to W^{-1,p'}(\Omega)$, defined by $(Au,v) = a(u,v)$, $\forall u,v \in W_0^{1,p}(\Omega)$, is monotone and demicontinuous (§ 2, Ch.I).

To enter in the frame given in § 1, we take $W = U$, a Hilbert space, $Z = X = Y = L^2(\Omega)$, $S : L^2(\Omega) \to L^2(\Omega)$ the identity operator, $F = B : U \to L^2(\Omega)$ linear, bounded and $M : L^2(\Omega) \to L^2(\Omega)$ is the realization of A in $L^2(\Omega)$. The maximality of M in $L^2(\Omega)$ follows by (5.5) which shows that A is coercive in $W_0^{1,p}(\Omega)$.

We define the regularizations of the mappings a_i by

(5.9) $$a_i^\varepsilon(y) = \int_{-\infty}^{\infty} a_i(y - \varepsilon\tau)\rho(\tau)d\tau, \qquad y \in R,$$

with ρ choosen as in § 3. An elementary calculation shows that the mappings a_i^ε satisfy (5.5)- (5.7) with modified constants, uniformly with respect to ε in a neighbourhood of the origin.

The family of operators M^ε, $\varepsilon > 0$, is obtained as M, starting from the functions a_i^ε.

We check the hypotheses (a), (b), (c).

Lemma 5.1. Let $f_\varepsilon \to f$ weakly in $L^2(Q)$. Then $y^\varepsilon = \theta_\varepsilon(f_\varepsilon) \to y = \theta(f)$ strongly in $C(0,T;L^2(\Omega))$, when $\varepsilon \to 0$.

Proof

Multiply by y^ε in (5.2), with a_i replaced by a_i^ε and Bu replaced by f_ε:

$$1/2\,|y^\varepsilon(t)|^2_{L^2(\Omega)} + \sum_{i=1}^N \int_0^t \int_\Omega a_i^\varepsilon(y^\varepsilon_{x_i}) y^\varepsilon_{x_i} = 1/2\,|y_0|^2_{L^2(\Omega)} + \int_0^t \int_\Omega f_\varepsilon\, y^\varepsilon.$$

From (5.5) we obtain that $\{y^\varepsilon\}$ is bounded in $L^\infty(0,T;L^2(\Omega)) \cap L^p(0,T;W_0^{1,p}(\Omega))$. Then (5.7) implies that $\{a_i^\varepsilon(y^\varepsilon_{x_i})\}$ is bounded in $L^{p'}(Q)$ and, from (5.2), $\{y_t^\varepsilon\}$ is bounded in $L^{p'}(0,T;W^{-1,p'}(\Omega))$. On a convenient subsequence, the Aubin theorem yields $y^\varepsilon \to y$ strongly in $L^p(0,T;L^2(\Omega))$.

We subtract two equations corresponding to $\varepsilon, \lambda > 0$ and multiply by $y^\varepsilon - y^\lambda$

$$1/2\,|y^\varepsilon(t) - y^\lambda(t)|^2_{L^2(\Omega)} + \eta \int_0^t \int_\Omega \sum_{i=1}^N |y^\varepsilon_{x_i} - y^\lambda_{x_i}|^2 \le \int_0^t \int_\Omega (f_\varepsilon - f_\lambda)(y^\varepsilon - y^\lambda).$$

Then $\{y^\varepsilon\}$ is a Cauchy sequence in $C(0,T;L^2(\Omega)) \cap L^2(0,T;H_0^1(\Omega))$. We show that $y = \theta(f)$. Let $v \in W_0^{1,p}(\Omega)$ be arbitrary fixed. We multiply by $y^\varepsilon - v$

$$1/2\; d/dt\,|y^\varepsilon(t) - v|^2_{L^2(\Omega)} + \sum_{i=1}^N \int_\Omega a_i^\varepsilon(y^\varepsilon_{x_i})(y^\varepsilon_{x_i} - v_{x_i})dx = \int_\Omega f_\varepsilon (y^\varepsilon - v)dx.$$

We integrate over the interval $[s,t] \subset [0,T]$, divide by $(t-s)$ and use the Schwartz inequality to deduce

$$((y^\varepsilon(t) - y^\varepsilon(s))/(t-s),\, y^\varepsilon(s) - v)_{L^2(\Omega)} + 1/(t-s) \sum_{i=1}^N \int_\Omega \int_s^t a_i^\varepsilon(v_{x_i})(y^\varepsilon_{x_i} - v_{x_i}) \le$$
$$\le 1/(t-s) \int_s^t \int_\Omega f_\varepsilon (y^\varepsilon - v).$$

Let $\varepsilon \to 0$

$$((y(t) - y(s))/(t-s),\, y(s) - v)_{L^2(\Omega)} + 1/(t-s) \sum_{i=1}^N \int_\Omega \int_s^t a_i(v_{x_i})(y_{x_i} - v_{x_i}) \le 1/(t-s) \int_s^t f(y - v).$$

Since y is differentiable with respect to t, in $W^{-1,p'}(\Omega)$, we can make $t \to s$ and we obtain the inequality:

$$(y_t(s), y(s) - v)_{W_0^{1,p}(\Omega) \times W^{-1,p'}(\Omega)} + \sum_{i=1}^N \int_\Omega a_i(v_{x_i})(y_{x_i}(s) - v_{x_i}) \le \int_\Omega f(s)(y(s) - v),$$

for any $v \in W_0^{1,p}(\Omega)$ and a.e. $s \in [0,T]$.

The operator A is maximal monotone in $W_0^{1,p}(\Omega) \times W^{-1,p'}(\Omega)$ and we obtain

$$y_t(s) - f(s) = \sum_{i=1}^{N} [a_i(y_{x_i}(s))]_{x_i} \qquad \text{a.e.}[0,T].$$

So $y = \theta(f)$ and, by the uniqueness of the solution, the convergence is true on the initial sequence.

Lemma 5.2. The mapping θ_ε is Gateaux differentiable in $L^2(Q)$ for any $\varepsilon > 0$. Moreover $r = \nabla\theta_\varepsilon(f)g$, with $f,g \in L^2(Q)$, satisfies

$$(5.10) \qquad r_t - \sum_{i=1}^{N} (a_i^\varepsilon(y_{x_i}^\varepsilon) r_{x_i})_{x_i} = g \qquad \text{in } Q,$$

$$(5.11) \qquad r(0,x) = 0 \qquad \text{in } \Omega,$$

$$(5.12) \qquad r(t,x) = 0 \qquad \text{in } \Sigma.$$

The solution of (5.10) - (5.12) is unique and $r \in L^\infty(0,T;L^2(\Omega)) \cap L^2(0,T;H_0^1(\Omega))$. Moreover $r_t \in L^s(0,T;W^{-1,s}(\Omega))$ for some $s > 1$.

Proof

Denote $y^\lambda = \theta_\varepsilon(f + \lambda g)$, $y = \theta_\varepsilon(f)$. We subtract the corresponding equations and we multiply by $y^\lambda - y$

$$(5.13) \qquad 1/2\,|y^\lambda(t) - y(t)|^2_{L^2(\Omega)} + \sum_{i=1}^{N} \int_\Omega \int_0^t |y^\lambda_{x_i} - y_{x_i}|^2 dx d\tau \le \lambda \int_0^t \int_\Omega g(y^\lambda - y) dx d\tau.$$

It yields that $z^\lambda = (y^\lambda - y)/\lambda$ is bounded in $L^\infty(0,T;L^2(\Omega))$ and in $L^2(0,T;H_0^1(\Omega))$ and $y^\lambda \to y$ strongly in $C(0,T;L^2(\Omega))$. The estimates for θ_ε show that $\{y^\lambda_{x_i}\}$ is bounded in $L^p(Q)$. By hypothesis (5.7) and the mean value theorem, we get

$$a_i^\varepsilon(y^\lambda_{x_i}) - a_i^\varepsilon(y_{x_i}) = m_i^\lambda \cdot (y^\lambda_{x_i} - y_{x_i}),$$

where m_i^λ depend on y^λ, y and are bounded in $L^{p/(p-2)}(Q)$.

Reconsidering (5.13) we see that

$$\sum_{i=1}^{N} \int_Q m_i^\lambda |z^\lambda_{x_i}|^2 dx dt$$

is bounded with respect to $\lambda > 0$.

Now we use the inequality of Young, generalizing the trick from Thm.3.8:

$$|m_i^\lambda z^\lambda_{x_i}| \le |m_i^\lambda|(1 + |z^\lambda_{x_i}|^{1+\nu}), \quad \nu > 0,$$

$$(5.14) \qquad |m_i^\lambda|\,|z^\lambda_{x_i}|^{1+\nu} = |m_i^\lambda|^{1-\mu}(|m_i^\lambda|^\mu \cdot |z^\lambda_{x_i}|^{1+\nu}) \le 1/\chi\, |m_i^\lambda|^{(1-\mu)\chi} + 1/\chi'\, |m_i^\lambda|^\mu \cdot |z^\lambda_{x_i}|^{(1+\nu)\chi'},$$

where we choose the constants $\mu = 1/2(1+\nu) < 1$, $1 < (1-\mu)\chi < p/(p-2)$, $\chi > 2$, $1/\chi + 1/\chi' = 1$.

The relation (5.14) implies the boundedness of $\{m_i^\lambda \cdot z^\lambda_{x_i}\}$ in $L^s(Q)$ for $s \in]1, p/(p-2)[$. On a convenient subsequence we have $z_t^\lambda \to r_t$ weakly in $L^s(0,T;W^{-1,s}(\Omega))$ and the Aubin theorem gives $z^\lambda \to r$ strongly in $L^2(Q)$ and weakly in $L^2(0,T;H_0^1(\Omega))$. Furthermore $(a_i^\varepsilon(y_{x_i}^\lambda) - (a_i^\varepsilon(y_{x_i}))/\lambda = m_i^\lambda z_{x_i}^\lambda \to h$ weakly in $L^s(Q)$. To identify the function h, we write

$$(a_i^\varepsilon(y_{x_i}^\lambda)) - (a_i^\varepsilon(y_{x_i}))/\lambda = [(a_i^\varepsilon(y_{x_i}^\lambda) - a_i^\varepsilon(y_{x_i}))/(y_{x_i}^\lambda - y_{x_i})] \cdot z_{x_i}^\lambda$$

and we have $z_{x_i}^\lambda \to r_{x_i}$ weakly in $L^2(Q)$. From $y_{x_i}^\lambda \to y_{x_i}$ strongly in $L^2(Q)$, the Egorov theorem states that for any $\sigma > 0$, there is $Q_\sigma \subset Q$, $mes(Q \setminus Q_\sigma) < \sigma$ and $y_{x_i}^\lambda \to y_{x_i}$ uniformly on Q_σ. The mappings a_i^ε are locally Lipschitzian, so

$$(a_i^\varepsilon(y_{x_i}^\lambda) - a_i^\varepsilon(y_{x_i}))/(y_{x_i}^\lambda - y_{x_i}) \to \dot{a}_i^\varepsilon(y_{x_i})$$

strongly in $L^2(Q_\sigma)$, for instance. We conclude that $h = \dot{a}_i^\varepsilon(y_{x_i})r_{x_i}$ a.e.Q and this ends the proof.

Let us consider the adjoint operator $\nabla\theta_\varepsilon(f)^* : L^2(Q) \to L^2(Q)$, $f \in L^2(Q)$. We denote $\nabla\theta_\varepsilon(f)^* q = p$ with $q, p \in L^2(Q)$ and we have that p satisfies:

(5.15) $\quad p_t + \sum_{i=1}^{N} [\dot{a}_i^\varepsilon(y_{x_i})p_{x_i}]_{x_i} = -q \quad$ in Q,

(5.16) $\quad p(T,x) = 0 \quad$ in Ω,

(5.17) $\quad p(t,x) = 0 \quad$ in Σ,

where $y = \theta_\varepsilon(f)$. This may be viewed by multiplying (5.15) with $r = \nabla\theta_\varepsilon(f)g$ and integrating by parts. The existence and the uniqueness of a generalized solution for the problem (5.15) - (5.17) are contained in the following non standard result for linear parabolic equations.

Theorem 5.3. The boundary value problem

(5.18) $\quad z_t - \sum_{i=1}^{N} [b_i(x,t)z_{x_i}]_{x_i} = v \quad$ in Q,

(5.19) $\quad z(t,x) = 0 \quad$ in Σ,

(5.20) $\quad z(0,x) = z_0(x) \quad$ in Ω,

where $b_i \in L^s(Q)$, $s > 1$ and

(5.21) $\quad b_i(t,x) \geq \eta > 0 \quad$ a.e.Q,

has a unique generalized solution $z \in L^2(0,T;H_0^1(\Omega))$ for $v \in L^2(Q)$, $z_0 \in L^2(\Omega)$. Moreover $z_t \in L^\tau(0,T;W^{-1,\tau}(\Omega))$ for some $\tau > 1$.

Here, a generalized solution is defined as an element $z \in L^2(0,T;H_0^1(\Omega))$, satisfying

$$\int_Q (-z\eta_t + \sum_{i=1}^{N} b_i z_{x_i} \eta_{x_i})dxdt = \int_\Omega z_0(x)\eta(x,0)dx + \int_Q f\eta\, dxdt,$$

for all $\eta \in H^1(Q)$ vanishing for $t = T$.

Proof

We consider the approximate equation

(5.22) $$z_t^\varepsilon - \sum_{i=1}^{N} [b_i^\varepsilon(x,t) z_{x_i}^\varepsilon]_{x_i} = v \text{ in } Q,$$

where b_i^ε are determined by

(5.23) $$b_i^\varepsilon(t,x) = \int_{R^{N+1}} b_i(t-\varepsilon\tau, x-\varepsilon y)\rho(\tau,y)d\tau dy,$$

and ρ is a Friedrichs mollifier as in §4.

The mappings b_i^ε satisfy (5.21) and

(5.24) $$b_i^\varepsilon \to b_i \text{ strongly in } L^s(Q),$$

(5.25) $$b_i^\varepsilon \in C^\infty(\bar{Q}).$$

It is wellknown that the problem (5.22), (5.20), (5.19) has a unique generalized solution (Ladyzhenskaya [66], Solonnikov [103]). Estimates of the type (5.13), (5.14) show that $\{z^\varepsilon\}$ is bounded in $L^\infty(0,T;L^2(\Omega)) \cap L^2(0,T;H_0^1(\Omega))$, $\{z_t^\varepsilon\}$ is bounded in $L^\tau(0,T;W^{-1,\tau}(\Omega))$, and $\{b_i^\varepsilon z_{x_i}^\varepsilon\}$ is bounded in $L^\tau(Q)$, $1 < \tau < s$. To pass to the limit, one may proceed as in Lemma 5.2.

The uniqueness is a direct consequence of (5.21).

Lemma 5.4. We denote $y^\varepsilon = \Theta_\varepsilon(f)$, $y = \Theta(f)$, $f \in L^2(Q)$ and we have

(5.26) $$|y^\varepsilon(t) - y(t)|_{L^2(\Omega)} \le C\varepsilon^{1/2}, \quad t \in [0,T],$$

with C independent of ε.

Proof

We subtract the equations for y^ε, y and we multiply by $y^\varepsilon - y$:

(5.27) $$1/2\, d/dt |y^\varepsilon(t) - y(t)|^2_{L^2(\Omega)} + \sum_{i=1}^{N} \int_\Omega (a_i^\varepsilon(y_{x_i}^\varepsilon) - a_i(y_{x_i}))(y_{x_i}^\varepsilon - y_{x_i}) = 0.$$

A simple computation gives

$$a_i^\varepsilon(y_{x_i}^\varepsilon) - a_i(y_{x_i}) = \int_{-1}^{1} m_i^\varepsilon \cdot (y_{x_i}^\varepsilon - \varepsilon\tau - y_{x_i})\rho(\tau)d\tau,$$

where the functions m_i^ε may be obtained by (5.17) and are bounded in $L^{p/(p-2)}(Q)$ with respect to ε in a neighbourhood of the origin.

We combine with (5.27) and integrate over $[0,t]$

$$1/2\,|y^{\varepsilon}(t)-y(t)|^2_{L^2(\Omega)}+\sum_{i=1}^{N}\int_0^t\iint_{\Omega}\int_{-1}^{1}m_i^{\varepsilon}\cdot(y^{\varepsilon}_{x_i}-\varepsilon\tau-y_{x_i})^2\rho(\tau)\;+$$

$$+\;\varepsilon\sum_{i=1}^{N}\int_0^t\iint_{\Omega}\int_{-1}^{1}m_i^{\varepsilon}\cdot(y^{\varepsilon}_{x_i}-\varepsilon\tau-y_{x_i})\,\tau\rho(\tau)=0.$$

From (5.6), $m_i^{\varepsilon}\geq\eta>0$ a.e. Q and the estimates (5.13), (5.14) achieve the proof.

We recall the approximate control problem:

(5.28) $$\text{Minimize}\left\{\int_0^T L^{\varepsilon}(y,u)dt+1/2\int_0^T|u-u^*|^2_U dt\right\},$$

(5.29) $$y_t-\sum_{i=1}^{N}(a_i^{\varepsilon}(y_{x_i}))_{x_i}=Bu \qquad \text{in } Q,$$

and (5.3), (5.4).The results of § 1 allow us to state:

Proposition 5.5. The problem (5.28) has solution $[y^{\varepsilon},u_{\varepsilon}]$ in $L^2(Q)\times L^2(0,T;U)$ and for any $\varepsilon>0$ there exists an adjoint arc $p^{\varepsilon}\in L^2(Q)$ such that

(5.30) $$p_t^{\varepsilon}+\sum_{i=1}^{N}[\dot a_i^{\varepsilon}(y^{\varepsilon}_{x_i})\cdot p^{\varepsilon}_{x_i}]_{x_i}=-q_{\varepsilon} \qquad \text{in } Q,$$

(5.31) $$p^{\varepsilon}(T,x)=0 \qquad \text{in } \Omega,$$

(5.32) $$p^{\varepsilon}(t,x)=0 \qquad \text{in } \Sigma,$$

(5.33) $$[q^{\varepsilon}(t),B^*p^{\varepsilon}(t)-u_{\varepsilon}(t)+u^*(t)]=\partial L^{\varepsilon}(y^{\varepsilon}(t),u_{\varepsilon}(t)) \qquad \text{in } [0,T].$$

Moreover $y^{\varepsilon}\to y^*$ strongly in $C(0,T;L^2(\Omega))$, $u_{\varepsilon}\to u^*$ strongly in $L^2(0,T;U)$, $p^{\varepsilon}\to p^*$ weakly* in $L^{\infty}(0,T;L^2(\Omega))$, $q_{\varepsilon}\to q^*$ weakly in $L^1(0,T;L^2(\Omega))$ and

(5.34) $$[q^*(t),B^*p^*(t)]\in\partial L(y^*(t),u^*(t)) \qquad \text{in } [0,T].$$

Remark 5.6. From the estimate (5.13) we see at once that $y^{\varepsilon}\to y^*$ strongly in $L^2(0,T;H_0^1(\Omega))$. Moreover, (5.5) gives $\{y^{\varepsilon}\}$ bounded in $L^p(0,T;W_0^{1,p}(\Omega))$.

To pass to the limit in the adjoint system we impose again the hypothesis (3.27) for the functions a_i.

Theorem 5.7. Let $[y^*,u^*]$ be an optimal pair for the problem (5.1) - (5.4). Then, there exist $p^*\in L^{\infty}(0,T;L^2(\Omega))\cap L^2(0,T;H_0^1(\Omega))$ and $q^*\in L^2(Q)$ which satisfy

$$p_t^*+\sum_{i=1}^{N}(Da_i(y^*_{x_i})p^*_{x_i})_{x_i}=-q^* \qquad \text{in } Q$$

and (5.31), (5.32), (5.34).

Proof

We multiply (5.30) by p^{ε} and we obtain the boundedness of $\{p^{\varepsilon}\}$ in $L^{\infty}(0,T;L^2(\Omega))\cap L^2(0,T;H_0^1(\Omega))$. Furthermore $\left\{\dot a_i^{\varepsilon}(y^{\varepsilon}_{x_i})\cdot|p^{\varepsilon}_{x_i}|^2\right\}$ is bounded in $L^1(Q)$. Since $\left\{\dot a_i^{\varepsilon}(y^{\varepsilon}_{x_i})\right\}$ is bounded in $L^{p/(p-2)}(Q)$, using again the inequality of Young, we get

$$a_i^{\varepsilon}(y^{\varepsilon}_{x_i})p^{\varepsilon}_{x_i}\to h^i \text{ weakly in } L^s(Q),$$

for $s\in]1, p/p-2[$. Moreover, the results of § 3.3, Ch.I, give $\dot{a}_i^{\varepsilon}(y^{\varepsilon}_{x_i}) \to \gamma^* \in Da_i(y^*_{x_i})$ weakly in $L^{p/(p-2)}(Q)$. We prove that

$$h^i \in Da_i(y^*_{x_i})p^*_{x_i} \qquad \text{a.e. } Q$$

by the hypothesis (3.27) and by the method of 3, based on the use of a concave-convex function.

Remark 5.8. If L is quadratic (the case without control constraints), from the maximum principle (5.34) we see that $u^* \in L^{\infty}(0,T;L^2(\Omega)) \cap L^2(0,T;H^1_0(\Omega))$. This regularity result is preserved under some constraints for the control, for instance if $u(t) \in U_{ad} = \{u \in L^2(\Omega);\ u \geq 0$ a.e. $\Omega\}$, a.e. $[0,T]$, as in § 2.

5.2 Delay systems

We continue in this section with the study of the problem

(5.35) $\quad$ Minimize $\int_0^T L(y,u)dt$,

(5.36) $\quad$ $y'(t) = A(y(t)) + Dy(t-h) + Eu(t) \qquad$ in $]0,T[$,

(5.37) $\quad$ $y(0) = y_0, \quad y(s) = \varphi(s) \qquad$ in $[-h,0]$.

We assume that $L : R^N \times R^m \to]-\infty,+\infty]$ is convex, lower semicontinuous, proper, with finite Hamiltonian. D,E are matrices with appropriate dimensions, $y_0 \in R^N$, $\varphi \in L^2(-h,0;R^N)$ and $A : R^N \to R^N$ is a Lipschitzian mapping.

Consider the spaces $Z = R^N \times L^2(-h,0;R^N)$, $X = R^N$, $W = R^m$, $Y = R^N \times \{0\}$. The operator $M : Z \to Z$ is given by

$$\mathrm{dom}(M) = \{[x,\varphi] \in R^N \times H^1(-h,0;R^N);\ \varphi(0) = x\},$$

$$M((x,\varphi) = [A\,x + D\varphi(-h),\ \dot{\varphi}\ ],$$

where $\dot{\varphi}$ is the derivative with respect to $s \in [-h,0]$. We preserve the notation φ' for the derivative with respect to $t \in [0,T]$.

The operator $F = [E,0]$ and S is the projection $S(y(t),y') = y(t)$. Here the pair $[y(t),y']$ plays the role of the state from § 1.

It is known that M is ω- maximal dissipative in Z, Badii and Webb [12]. Then the equation (5.36), (5.37) has a unique solution $y \in C(0,T;R^N)$ such that $y' \in L^2(0,T;R^N)$. This may be also seen directly, by integrating (5.36), (5.37) on intervals of length h, successively.

The operators M^{ε} are obtained similarly, by replacing A with the regularization A^{ε}

(5.38) $\quad$ $A^{\varepsilon}(y) = \int_{R^N} A(y - \varepsilon x)\rho(x)dx, \quad \varepsilon > 0, \quad y \in R^N,$

where ρ is a Friedrichs mollifier as in § 4. The family of mappings $A^{\varepsilon} : R^N \to R^N$ is Lipschitzian, uniformly with respect to ε.

Obviously, θ and θ_{ε} are well defined. The approximate control problem is

(5.39) $\qquad \text{Minimize}\left\{\int_0^T L^{\varepsilon}(y,u)dt + 1/2\,|u - u^*|^2_{L^2(0,T;R^m)}\right\},$

(5.40) $\qquad y'(t) = A^{\varepsilon}(y(t)) + Dy(t-h) + Eu(t) \qquad \text{in } [0,T]$

and (5.37).

We outline the checking of the conditions (a), (b), (c) of § 1.

Let $u_\varepsilon \to u$ weakly in $L^2(0,T;R^m)$. We multiply (5.40) by $y_\varepsilon = \theta_\varepsilon(Eu_\varepsilon)$ and integrate

$$1/2\,|y_\varepsilon(t)|^2 - 1/2\,|y_0|^2 = \int_0^t A^{\varepsilon}(y_\varepsilon(t))y_\varepsilon(t)dt + \int_0^t Dy_\varepsilon(t-h)y_\varepsilon(t)dt + \int_0^t Eu_\varepsilon(t)y_\varepsilon(t)dt.$$

Here $|.|$ is the Euclidean norm. As A^{ε} are uniformly Lipschitzian with respect to ε, the Gronwall lemma shows that $\{y_\varepsilon\}$ is bounded in $L^\infty(0,T;R^N)$ and $\{A^{\varepsilon}(y_\varepsilon)\}$ is bounded in $L^\infty(0,T;R^N)$. Then (5.40) implies that $\{y'_\varepsilon\}$ is bounded in $L^2(0,T;R^N)$, that is $y_\varepsilon \to y$ strongly in $C(0,T;R^N)$, $y'_\varepsilon \to y'$ weakly in $L^2(0,T;R^N)$.

To pass to the limit in (5.40) we remark that:

$$A^{\varepsilon}(y_\varepsilon(t)) - A(y(t)) = \int_{R^N} [A(y_\varepsilon(t) - \varepsilon x) - A(y(t))]\rho(x)dx \to 0$$

as $\varepsilon \to 0$. Therefore $y = \theta(Eu)$ and (a) is proved.

It is easy to see that θ_ε is Gateaux differentiable and, for $w,v \in L^2(0,T;R^N)$, $r = \nabla\theta_\varepsilon(w)v$ satisfies

(5.41) $\qquad r'(t) = \text{grad}\, A^{\varepsilon}(\theta_\varepsilon(w))r(t) + Dr(t-h) + v(t), \qquad \text{in } [0,T],$

(5.42) $\qquad r(0) = 0,\ r(s) = 0, \qquad \text{on } [-h,0],$

where $\text{grad}\, A^{\varepsilon}$ is the Jacobian matrix of A^{ε}.

The adjoint operator $\nabla\theta_\varepsilon(w)^* : L^2(0,T;R^N) \to L^2(0,T;R^N)$ is given by $\nabla\theta_\varepsilon(w)^*q = p$, for $w,q,p \in L^2(0,T;R^N)$ and

(5.43) $\qquad -p'(t) = \text{grad}\, A^{\varepsilon}(\theta_\varepsilon(w))^*p(t) + D^*p(t+h) + q(t),$

(5.44) $\qquad p(T) = 0,\ p(s) = 0 \qquad \text{in } [T,T+h].$

We have denoted by D^*, respectively $\text{grad}\, A^{\varepsilon}(.)^*$ the transposed matrices.

The conditions (c) may be immediately checked with $\delta(\varepsilon) = \varepsilon$, so, in this case, we have $L^{\varepsilon} = L_\varepsilon$, the Yosida regularization of order ε of the convex mapping L.

Let $[y_\varepsilon, u_\varepsilon]$ be an optimal solution of the problem (5.39), (5.40), (5.37). The approximate optimality system, given by § 1, is

(5.45) $\qquad y'_\varepsilon = A^{\varepsilon}(y_\varepsilon(t)) + Dy_\varepsilon(t-h) + Eu_\varepsilon(t) \qquad \text{in } [0,T],$

(5.46) $\qquad -p'_\varepsilon = \text{grad}\, A^{\varepsilon}(y_\varepsilon(t))^*p_\varepsilon(t) + D^*p_\varepsilon(t+h) - q_\varepsilon(t) \text{ in } [0,T],$

(5.47) $\qquad y_\varepsilon(0) = y_0,\ y_\varepsilon(s) = \varphi(s) \qquad \text{in } [-h,0]$

(5.48) $\qquad p_\varepsilon(T) = 0,\ \ p_\varepsilon(s) = 0, \qquad \text{in } [T,\ T+h],$

(5.49) $[q_\varepsilon(t), E^* p_\varepsilon(t) - u_\varepsilon(t) + u^*(t)] = \partial L_\varepsilon(y_\varepsilon(t), u_\varepsilon(t))$ in $[0,T]$.

Moreover, we know that $y_\varepsilon \to y^*$ strongly in $C(0,T;R^N)$, $y_\varepsilon' \to (y^*)'$ strongly in $L^2(0,T;R^N)$, $u_\varepsilon \to u^*$ strongly in $L^2(0,T;R^m)$, $p_\varepsilon \to p^*$ weakly* in $L^\infty(0,T;R^N)$, $q_\varepsilon \to q^*$ weakly in $L^1(0,T;R^N)$ and

(5.50) $[q^*(t), E^* p^*(t)] \in \partial L(y^*(t), u^*(t))$ in $[0,T]$.

By (5.45) we get $p_\varepsilon' \to (p^*)'$ weakly in $L^1(0,T;R^N)$ since $\{\text{grad } A^\varepsilon(y_\varepsilon)^*\}$ is bounded in $L^\infty(0,T;R^{NxN})$ and $q_\varepsilon \to q^*$ weakly in $L^1(0,T;R^N)$.

The Helly compactness principle, combined with the dominated convergence theorem of Lebesgue, implies $p_\varepsilon \to p^*$ strongly in $L^2(0,T;R^N)$, for instance.

Since $\{\text{grad } A^\varepsilon(y_\varepsilon)^*\}$ is bounded in $L^\infty(0,T;R^{NxN})$, by a variant of Thm.3.14, Ch.I, we have

$$\text{grad } A^\varepsilon(y_\varepsilon)^* \to DA(y^*)^*$$

weakly* in $L^\infty(0,T;R^N)$, where $DA(y^*)$ is the Clarke generalized gradient associated to the vector Lipschitzian mapping A.

We have proved the theorem

Theorem 5.9. *Let $[y^*,u^*]$ be an optimal pair for the problem* (5.35) - (5.37). *There exist* $p^* \in W^{1,1}(0,T;R^N)$, $q^* \in L^1(0,T;R^N)$ *such that*

$$(y^*)'(t) = A(y^*(t)) + Dy^*(t-h) + Eu^*(t) \quad \text{in } [0,T],$$

$$-(p^*)'(t) = DA(y^*(t))^* p^*(t) + D^* p^*(t+h) - g^*(t) \quad \text{in } [0,T],$$

$$y^*(0) = y_0, \quad y^*(s) = \varphi(s) \quad \text{in } [-h,0],$$

$$p^*(T) = 0, \quad p^*(s) = 0 \quad \text{in } [T,T+h],$$

$$[q^*(t), E^* p(t)] \in \partial L(y^*(t), u^*(t)) \quad \text{in } [0,T].$$

Moreover, the functions y^*,u^*,p^*,q^* *are limits of the functions* $y_\varepsilon, u_\varepsilon, p_\varepsilon, q_\varepsilon$, *solutions of the approximate optimality system* (5.44) - (5.48).

Remark 5.10. If no control constraints are imposed and L is quadratic, by (5.49) we see that $u^* \in W^{1,2}(0,T;R^m)$. This regularity result is obviously preserved under positivity constraints on the control.

III. VARIATIONAL INEQUALITIES

The problems we study in this chapter in §1, §2 can be formally written as those from Chapter II, § 2, § 3. The essential difference is that, now, the nonlinear term $\beta(.)$ is the subdifferential of a convex, lower semicontinuous, proper function and may be neither locally Lipschitzian, nor satisfy the growth condition (2.5). The abstract scheme from § 1, Chapter II, and the general smoothing procedure are used similarly, but, in order to pass to the limit in the approximate adjoint equation, different arguments are needed.

In §3 we discuss the difficult problem of the optimal control for the vibrating string with obstacle, in the setting of singular control systems, according to § 4, Chapter II.

Section 4 is devoted to the analysis of the relationship between the control problems governed by variational inequalities and the problems with state constraints, which leads to efficient approximation methods and to bang-bang results for the optimal control. The large area of application of this approach is justified in the next section by the investigation of an optimal design problem with unilateral conditions and state constraints.

Finally, we examine an optimal control problem governed by a semilinear elliptic variational inequality and we put into evidence another useful method.

1. Parabolic problems

We consider the problem:

(1.1) Minimize $\int_0^T L(y,u)dt$,

(1.2) $y_t - \Delta y + \beta(y) \ni Fu$ in Q,

(1.3) $y(0,x) = y_0(x)$ in Ω,

(1.4) $y(t,x) = 0$ in Σ.

Here $\beta \subset R \times R$ is a maximal monotone graph, possibly multivalued, so it doesn't satisfy (2.5), Chapter II. The other assumptions from § 2, II are maintained, that is $F : W \to L^2(\Omega)$ is a linear, continuous operator, $L : L^2(\Omega) \times W \to]-\infty,+\infty]$ is a convex, lower semicontinuous, proper function, with finite Hamiltonian and $y_0 \in H_0^1(\Omega)$, $y_0(x) \in \text{dom}\beta$ a.e. in Ω. Then, the variational inequality (1.2) with the mixed conditions (1.3), (1.4) has a unique solution $y \in L^2(0,T;H^2(\Omega) \cap H_0^1(\Omega))$, $y_t \in L^2(Q)$, according to Thm.4.5, Ch.I.

We present several examples to clarify the significance of the problem (1.2) - (1.4):

Example 1. The obstacle problem.

Let β be given by

$$(1.5)\qquad \beta(y) = \begin{cases} 0 & y > 0 \\ ]-\infty,0] & y = 0, \\ \emptyset & y > 0. \end{cases}$$

Then (1.2) may be equivalently rewritten as

$$\begin{array}{ll} y \geq 0 & \text{a.e.}Q, \\ y_t - \Delta y - Fu \geq 0 & \text{a.e.}Q, \\ y(y_t - \Delta y - Fu) = 0 & \text{a.e.}Q. \end{array}$$

If y measures the evolution of the temperature in the domain Ω and in the period [0,T], then (1.2) may be interpreted as the model of the action of a thermostat. The subset of Q for which $y(t,x) = 0$ is called the coincidence set. Its complementary in Q is called the noncoincidence set. Their common boundary has the name of free boundary since it is not a priori known and has an evolution in time. A direct generalization of this example is obtained when the "obstacle" depends on $(x,t)\in Q$, that is β is a function of $(x,t)\in Q$ too:

$$\beta(x,t,y) = \begin{cases} 0 & y > O(x,t), \\ ]-\infty,0] & y = O(x,t), \\ \emptyset & y < O(x,t). \end{cases}$$

For regular mappings O and usual compatibility conditions ($O(x,t) \leq 0$ on Σ , $O(x,0) \leq y_0(x)$ in Ω), Thm.4.5, Ch.I may be applied in this case too.

Example 2. The problem with two obstacles.

Now, we consider β of the form

$$(1.6)\qquad \beta(y) = \begin{cases} 0 & a < y < b, \\ ]-\infty,0] & y = a, \\ [0,+\infty] & y = b, \\ \emptyset & y < a \text{ or } y > b, \end{cases}$$

where a,b are two given real constants with $0\in[a,b]$ (compatibility with the Dirichlet conditions). We suppose that $y_0(x)\in[a,b]$ a.e.Ω. Then (1.2) may be equivalently rewritten as

$$\begin{array}{ll} a \leq y \leq b & \text{in } Q, \\ y_t - \Delta y - Fu \geq 0 & \text{if } y = a, \\ y_t - \Delta y - Fu \leq 0 & \text{if } y = b, \\ y_t - \Delta y = Fu & \text{if } a < y < b. \end{array}$$

All the commments from the previous example may be extended to this one.

Example 3. The one-phase Stefan problem.

The notion of "free boundary" has been introduced in Example 1 in a rather artificial manner. Now, we shall see the natural meaning and importance of this concept, which characterizes a remarkable class of nonlinear problems, just called "free boundary problems".

The example we discuss has the simplest possible structure and we are mainly interested in the relationship with variational inequalities. See Chapter IV for a detailed analysis of other types of problems from this category.

Let $G \subset \Omega$ be a subdomain. We assume that at the moment $t = 0$, G contains ice at the temperature 0°, and $\Omega - G$ is formed of water. At the boundary of Ω, we preserve a given positive temperature O_1. As a result of the heat transfer, the ice continues to melt, remaining at 0° in the interior. The area occupied by the ice decreases and we denote it by $G(t)$. The surface of separation between ice and water, $\partial G(t)$, is the free boundary and now it has a precise physical meaning. We make the hypothesis that $\partial G(t)$ is described by the equation $t = l(x)$. In the region filled with water, the temperature O satisfies an equation of the type (with normalized coefficients):

$$O_t(x,t) - \Delta O(x,t) = 0 \quad \text{in } (x,t) \in Q,\ l(x) < t < T,$$
$$O(x,t) = O_1(x,t) \quad \text{in } \Sigma.$$

In $G(t)$ we have $O(x,t) = 0$ and on the free boundary we impose the Stefan condition for the conservation of energy

$$\text{grad } O(t,x) \,.\, \text{grad } l(x) = -L, \text{ in } (x,t) \in Q,\ \ t = l(x),$$

where L is a given positive constant.

For simplicity, we take $G = G(0) = \Omega$, so we have the initial condition $O(x,0) = 0$ in Ω. We denote $y(t,x) = \int_0^t O(x,\tau)d\tau, x \in \Omega, t \in [0,T]$. If $t \leq l(x)$, $y(x,t) = 0$, and if $t > l(x)$

$$y(x,t) = \int_{l(x)}^t O(x,\tau)d\tau.$$

Let $t > l(x)$. Then:

$$y_{x_i}(x,t) = \int_{l(x)}^t O_{x_i}(x,\tau)d\tau - l_{x_i}(x)O(x,l(x)) = \int_{l(x)}^t O_{x_i}(x,\tau)d\tau;$$
$$y_{x_i x_i}(x,t) = \int_{l(x)}^t O_{x_i x_i}(x,\tau)d\tau - l_{x_i}(x)O_{x_i}(x,l(x)).$$

It yields that, for $t > l(x)$, we have

$$\Delta y(t,x) = \int_{l(x)}^t \Delta O(x,\tau)d\tau + L = y_t(x,t) + L.$$

Consequently, y satisfies to

$$y_t - \Delta y = -L \qquad \text{if } y > 0,$$
$$y_t - \Delta y > -L \qquad \text{if } y = 0.$$

We see that y is the solution of the variational inequality

$$y_t - \Delta y + \beta(y) = -L \qquad \text{in } Q,$$
$$y(x,0) = 0 \qquad \text{in } \Omega,$$
$$y(x,t) = \int_0^t \theta_1(x,\tau)d\tau \qquad \text{in } \Sigma,$$

where β is given by (1.5).

Now, the relationship between the abstract definition of the free boundary from Example 1 and its physical support is clear.

Remark 1.1. The transformation which relates y and θ was introduced in free boundary problems by Baiocchi [25], and it was applied to one-phase Stefan problems by Duvaut [41] and Fremond [53]. For a general investigation of the Stefan problem we quote the works of Ockendon and Elliott [45], Friedman [47], Rubenstein [100].

Now, we return to the control problem (1.1)-(1.4). We apply the usual smoothing technique and we define

(1.7) $$\beta^\varepsilon(y) = \int_{-\infty}^{\infty} \beta_\varepsilon(y - \varepsilon\tau)\rho(\tau)d\tau$$

with β_ε the Yosida approximation of β and ρ as in Chapter II, § 2. We denote by $[y^*, u^*]$ an optimal pair for the problem (1.1)-(1.4) and by $[y^\varepsilon, u_\varepsilon]$ an optimal pair for the approximate problem

(1.8) $$\text{Minimize}\left\{\int_0^T L^\varepsilon(y,u)dt + 1/2 \int_0^T |u - u^*|_W^2 dt\right\},$$

(1.9) $$y_t - \Delta y + \beta^\varepsilon(y) = Fu \qquad \text{in } Q,$$

and (1.3), (1.4). Here L^ε is defined in § 1, Chapter II, and the existence of $[y^*,u^*]$, $[y^\varepsilon,u_\varepsilon]$ follows easily under appropriate coercivity assumptions.

By the results of § 1, II and similarly to § 2, II, we obtain

<u>Proposition 1.2. There is</u> $p^\varepsilon \in C(0,T;L^2(\Omega))$ <u>which satisfies the approximate optimality system</u>

(1.10) $$p_t^\varepsilon + \Delta p^\varepsilon - \dot\beta^\varepsilon(y^\varepsilon)p^\varepsilon = \partial_1 L^\varepsilon(y^\varepsilon, u_\varepsilon) \qquad \text{in } Q,$$
$$p^\varepsilon(T,x) = 0 \quad \text{in } \Omega, \quad p^\varepsilon(t,x) = 0 \qquad \text{in } \Sigma,$$

(1.11) $$F^* p^\varepsilon = \partial_2 L^\varepsilon(y^\varepsilon, u_\varepsilon) + u_\varepsilon - u^* \qquad \text{in } [0,T],$$

together with $y^\varepsilon, u_\varepsilon$. Moreover $u_\varepsilon \to u^*$ strongly in $L^2(0,T;W)$, $y^\varepsilon \to y^*$ strongly in $C(0,T;L^2(\Omega))$,

$\partial_1 L^\varepsilon(y^\varepsilon, u_\varepsilon) \to q^*$ weakly in $L^1(0,T;L^2(\Omega))$, $p^\varepsilon \to p^*$ weakly in $L^2(0,T;H_0^1(\Omega))$ and

$$(1.12) \qquad [q^*, F^* p^*] \in \partial L(y^*, u^*) \qquad \text{in } [0,T].$$

Remark 1.3. If $W = L^2(\Omega)$ and we have no control constraints (L is a quadratic functional), by (1.12) we get $u^* \in L^2(0,T;H^1(\Omega))$.

Since the condition (2.5), Chapter II, is not fulfilled, to pass to the limit in (1.10) and to obtain more information on p^*, we make use of the structure properties of β, according to Example 2, §3.1, Ch.I.

Let β^0 be the minimal section of β, that is $\beta^0(y) \in \beta(y)$ and $|\beta^0(y)| = \inf\{|z|;\ z \in \beta(y)\}$; β^0 is a nondecreasing function on domβ and we write $\beta^0 = \tilde{\beta}_1 + \beta_2$, where $\tilde{\beta}_1$ is the "jump" mapping associated to β^0 and $\beta_2 = \beta^0 - \tilde{\beta}_1$ is a nondecreasing, continuous function on domβ. Then we have $\beta = \beta_1 + \beta_2$, where β_1 is the maximal monotone operator generated by $\tilde{\beta}_1$, that is $\beta_1(y) = [\tilde{\beta}_1(y-), \tilde{\beta}_1(y+)]$ on intdomβ, etc., as in Example 2, §3.1, I.

Now, we are able to impose the hypotheses on β. We ask that $\tilde{\beta}_1$ has a finite number of discontinuities and we denote by D their set; for β_2 we require to be locally Lipschitzian and to satisfy the condition (3.27) from Chapter II.

The equation (1.10) may be rewritten as

$$(1.13) \qquad p_t^\varepsilon + \Delta p^\varepsilon - \dot{\beta}_1^\varepsilon(y^\varepsilon)p^\varepsilon - \dot{\beta}_2^\varepsilon(y^\varepsilon)p^\varepsilon = \partial_1 L^\varepsilon(y^\varepsilon, u_\varepsilon),$$

and $\beta_1^\varepsilon, \beta_2^\varepsilon$ are obtained from (1.7) by:

$$(1.14) \qquad \beta_2^\varepsilon(y) = \int_{-\infty}^{\infty} \beta_2((I + \varepsilon\beta)^{-1}(y - \varepsilon\tau))\rho(\tau)d\tau,$$

$$(1.15) \qquad \beta_1^\varepsilon(y) = \beta^\varepsilon(y) - \beta_2^\varepsilon(y).$$

We assume, for simplicity, that Ω is a domain in R.

Theorem 1.4. Under the above conditions, there is $\delta \in \mathcal{D}'(Q)$ with supp$\delta \subset Q^* = \{(t,x) \in Q;\ y^*(t,x) \in D\}$ such that:

$$(1.16) \qquad p_t^* + \Delta p^* - \delta - D\beta_2(y^*)p^* \ni q^* \qquad \text{in } Q.$$

Moreover, if $\{\partial_1 L^\varepsilon(y^\varepsilon, u_\varepsilon)\}$ is bounded in $L^2(Q)$, then $p^\varepsilon \to p^*$ uniformly on compact subsets of $Q \setminus Q^*$.

Proof

Since $\{u_\varepsilon\}$ is bounded in $L^2(0,T;W)$, by (2.6) Ch.II, we see that $\{y^\varepsilon\}$ is bounded in $L^2(0,T;H^2(\Omega) \cap H_0^1(\Omega))$ and $\{y_t^\varepsilon\}$ is bounded in $L^2(Q)$. The Sobolev imbedding theorem, as $\Omega \subset R$, gives $y^\varepsilon \to y^*$ uniformly on $\bar{Q}$ and belong to $C(\bar{Q})$. Obviously, from β_2 locally Lipschitzian, we have $\{\dot{\beta}_2^\varepsilon(y^\varepsilon)\}$ bounded in $L^\infty(Q)$. By Thm.3.14, Chapter I, it yields

$$\dot{\beta}_2^\varepsilon(y^\varepsilon) \to D\beta_2(y^*)$$

weakly* in $L^\infty(Q)$. A simplified variant of the argument from § 3, Ch. II based on the use of a saddle function, shows that, under condition (3.27) Ch. II, we have

$$\dot{\beta}_2^\varepsilon(y^\varepsilon)p^\varepsilon \to D\beta_2(y^*)p^*$$

weakly in $L^2(Q)$. On the other hand $Q \setminus Q^*$ is an open subset, not necessarily connected. Let $Q \setminus Q^* = \bigcup_{i=1}^{m} Q_i$, where $Q_i \subset Q$ are open subsets given by

$$Q_i = \left\{ (t,x) \in Q;\ d_i < y^*(t,x) < d_{i+1} \right\},$$

and $\left\{d_i\right\}_{i=\overline{1,m}} = D$.

Let $K_i \subset Q_i$ be compact subsets, $K = \bigcup_{i=1}^{m} K_i$, and $d > 0$ be sufficiently small.

For $\varepsilon \le \varepsilon_0$, ε_0 sufficiently small, the range of $y^\varepsilon(t,x)$, $(t,x) \in K$, is disjoint from a neighbourhood $\tilde{D}$ of D, obtained as the union of intervals of length 2d, centered in points of D.

But, from the structure of β_1, if ε is small, then β_1^ε approximates β_1 as graphs in R x R and, in particular, $\beta_1^\varepsilon = \beta_1$ in the exterior of $\tilde{D}$. Therefore, for $(t,x) \in K$, we have $\dot{\beta}_1^\varepsilon(y^\varepsilon(t,x)) = 0$.

We remark that, from (1.13) and the above estimates, we have $\dot{\beta}_1^\varepsilon(y^\varepsilon)p^\varepsilon \to \delta$ in $\mathcal{D}'(Q)$. The above discussion shows that δ vanishes on $Q \setminus Q^*$.

To justify the last statement of the theorem, let us take (to fix the ideas) K_1 to be a compact subset of Q_1. For ε sufficiently small, the equation (1.13) becomes on K_1:

$$p_t^\varepsilon + \Delta p^\varepsilon - \dot{\beta}_2^\varepsilon(y^\varepsilon)p^\varepsilon = \partial_1 L^\varepsilon(y^\varepsilon, u_\varepsilon).$$

Let $\tau \in \mathcal{D}(Q \setminus Q^*)$, $\tau = 1$ on K_1. We denote by v^ε the extension of τp^ε, by 0, to R x]0,T[. It is easy to see that v^ε is a solution of the Cauchy problem:

$$\begin{aligned} & v_t^\varepsilon + \Delta v^\varepsilon = f_\varepsilon && \text{in }]0,T[\times R, \\ & v\ (T,x) = 0 && \text{in } R, \end{aligned}$$

where $f_\varepsilon = \tau(\partial_1 L^\varepsilon(y^\varepsilon, u_\varepsilon) + \dot{\beta}_2^\varepsilon(y^\varepsilon)p^\varepsilon) + p^\varepsilon(\tau_t + \Delta\tau) + 2\,\mathrm{grad}\,\tau.\mathrm{grad}\, p^\varepsilon$ and it is bounded in $L^2([0,T] \times R)$.

Standard estimates show that $\left\{v_t^\varepsilon\right\}$, $\left\{\Delta v^\varepsilon\right\}$ are bounded in $L^2([0,T] \times R)$. Since $\tau = 1$ on K_1, we obtain $\left\{p^\varepsilon\right\}$ bounded in $W^{2,1,2}(K_1)$ and the Sobolev embedding theorem finishes the proof.

Remark 1.5. The condition R is not essential, according to Remark 2.7, Chapter II. See also Thm.4.5, Chapter I.

Remark 1.6. The optimality conditions for control problems governed by parabolic variational inequalities are analysed, in a different form, in the book of V.Barbu [13]. Recently, using a variant of the adapted penalization method, Zheng - Xu He [56] has obtained necessary optimality conditions for parabolic problems with state constraints, both in the case of variational inequalities and in the case of semilinear equations. The above result unifies both situations, and, in § 4, we also discuss the state constrained problem.

Let $W = L^2(\Omega)$ and $u \in U_{ad} = \{u \in L^2(\Omega);\ 0 \le u(x) \le 1 \text{ a.e. } \Omega\}$. Then $\{u_\varepsilon\}$ is bounded in $L^p(Q)$ $p \ge 2$, and the result of Thm.1.4 is valid for $\Omega \subset R^N$ according to Remark 1.5.

Let j be the indicator function of the interval [0,1] and $\beta(y) = \partial j(y) + y$.

Corollary 1.7. There is $p^* \in L^2(0,T;H_0^1(\Omega))$ and $\delta \in \mathcal{D}'(Q)$ with $\operatorname{supp}\delta \subset Q^* = \{(t,x) \in Q;\ y^*(t,x) = 0 \text{ or } y^*(t,x) = 1\}$, such that

$$p_t^* + \Delta p^* - \delta + p^* = q^* \text{ in } Q. \tag{1.17}$$

This is just a special case of (1.16).

We choose F to be the identity in $L^2(\Omega)$ and $L(y,u) = 1/2\,|y|^2_{L^2(\Omega)} + I_{U_{ad}}(u)$, where

$$I_{U_{ad}}(u) = \begin{cases} 0 & u(t,.) \in U_{ad} \quad \text{a.e.}[0,T], \\ +\infty & \text{otherwise.} \end{cases}$$

We remark that the solution of this simplified control problem remains nontrivial due to the generality of the initial condition y_0.

It yields that $q^* = y^*$ and (1.12) is equivalent with

$$\int_Q p^*(v - u^*) \le 0, \qquad v \in \tilde{U}_{ad} \tag{1.18}$$

$$\tilde{U}_{ad} = \{u \in L^2(Q);\ u(t,.) \in U_{ad} \text{ a.e.}[0,T]\}.$$

Take $v = u^*$ on Q^*; we get

$$\int_{Q-Q^*} p^*(v - u^*) \le 0.$$

We denote $Q_1 = \{(t,x) \in Q - Q^*;\ p^*(t,x) > 0\}$ and $Q_2 = \{(t,x) \in Q - Q^*;\ p^*(t,x) < 0\}$ and we obtain at once that $u^* = 0$ on Q_2 and $u^* = 1$ on Q_1.

By Thm.4.5, Chapter I and by the choice of U_{ad}, we see that $y^* \in W^{2,1,p}(Q)$ for any $p \ge 2$, therefore $q^* \in W^{2,1,p}(Q)$. On $Q - Q^*$ we have $\delta = 0$ and the argument from the end of

Thm.1.4 shows that $p^* \in W^{2,1,p}_{loc}(Q - Q^*)$, any $p > 2$.

On $Q_3 = Q - (Q_1 \cup Q_2 \cup Q^*)$ we have $p^* = 0$. If meas $Q_3 > 0$, by (1.17) and the regularity of p^*, it follows $y^* = 0$. This contradicts the definition of Q^* and we conclude that u^* is a bang-bang control on Q Q^*.

On Q^*, by (1.2), we have $u^* \in \beta(y^*)$ a.e. Q^*.

However $\beta(y^*) \leq 0$ for $y^* = 0$ and $u^* \geq 0$ by the definition of U_{ad}. It yields that $u^* = 0$ on the set $y^* = 0$. Similarly, we remark that $\beta(y^*) \geq 1$ for $y^* = 1$ and $u^* \leq 1$ from the definition of U_{ad}. So $u^* = 1$ on the set where $y^* = 1$.

We have established.

Corollary 1.8. Under the above conditions, any optimal control is bang-bang on Q.

Remark 1.9. Bang-bang results for optimal control problems governed by elliptic or parabolic variational inequalities or by free boundary problems were obtained by A.Friedman [49], [50], A.Friedman and D.Yaniro [51], A.Friedman and L.Jiang [48]. We also quote the results of V.Barbu [17] for elliptic variational inequalities and of G.Morosanu and Zheng - Xu He [57] for variational inequalities associated with the biharmonic operator.

Remark 1.10. In the case without control constraints, necessary optimality conditions for control problems governed by parabolic variational inequalities were obtained by F.Mignot and J.P.Puel [75] by means of the conical derivative, according to Mignot [74]. Similar results are obtained in the works of C.Saguez and A.Bermudez [108], F.Mignot and J.P.Puel [76], by using an equivalence relation between the problems governed by variational inequalities and certain control problems with state constraints. Another comparable approach will be analysed in § 4.

2. Hyperbolic problems

In this section we complete the study from § 3, Chapter II, with the analysis of the case when β is unbounded, i.e. β is a maximal monotone operator, possibly multivalued.

Since the available regularity results are less powerful in the hyperbolic case as compared with the parabolic situation, we limit ourselves to the discussion of the following problem

(2.1) $\quad$ Minimize $\int_0^T L(y,u)dt$,

(2.2) $\quad y_{tt} - \Delta y + \beta(y_t) \ni Bu \quad$ in Q,

(2.3) $\quad y(0,x) = y_0(x),\ y_t(0,x) = v_0(x) \quad$ in Ω,

(2.4) $\quad y(t,x) = 0 \quad$ in Σ,

where β is a maximal monotone graph in R x R.

The approximation procedure of § 3.1, Ch.II, applies identically here. Using the same notations, Proposition 3.4, II, remains valid. In order to pass to the limit in the adjoint system we impose to β the same conditions as in § 1, that is $\beta = \beta_1 + \beta_2$, where β_2 is locally Lipschitzian satisfying (3.27) and β_1 has a finite number of discontinuities on domβ.

Theorem 2.1. There is $m^* \in W^{1,\infty}(0,T;L^2(\Omega)) \cap L^\infty(0,T;H_0^1(\Omega))$ and $\delta \in \mathcal{D}'(Q)$ such that

(2.5) $m^*_{tt} - \Delta m^* + \delta = -\int_t^T q^*$ in Q,

(2.6) $[q^*(t), -B^* m^*_t(t)] \in \partial L(y^*(t), u^*(t))$ in $[0,T]$.

Let $Q_0 = \{(t,x) \in Q;\ y^*_t(t,x) \notin D\}$. Then

(2.7) $\dot{\beta}^\varepsilon_1(y^\varepsilon_t) m^\varepsilon_t \to 0$ a.e. Q_0.

Moreover, if $\Omega \subset R$, then $\delta = \delta_1 + D\beta_2(y^*_t) \cdot m^*_t$ and $\operatorname{supp}\delta_1 \subset Q \setminus Q_0$.

Proof

By Proposition 3.4, Ch. II we know that $m^\varepsilon \to m^*$ weakly* in $L^\infty(0,T;H_0^1(\Omega)) \cap W^{1,\infty}(0,T;L^2(\Omega))$ and $q_\varepsilon \to q^*$ weakly in $L^1(0,T;L^2(\Omega))$. We denote by $\delta \in \mathcal{D}'(Q)$ the distributional limit of $\dot{\beta}^\varepsilon(y^\varepsilon_t) m^\varepsilon_t$ and we obtain (2.5). The relation (2.6) coincides with (3.23), Ch. II.

Suppose now that mes $Q_0 > 0$, otherwise (2.7) is trivial. We have $D = \{d_i\}_{i=\overline{1,m}}$ and $Q_0 = \bigcup_{i=1}^m Q_i$, $Q_i = \{(t,x) \in Q;\ d_i < y^*_t(t,x) < d_{i+1}\}$. We may assume that mes $Q_i > 0$ without loss of generality. We fix our attention on Q_1 and we can write it as

$$Q_1 = \bigcup_{c>0} Q_1^c$$

where $Q_1^c = \{(t,x) \in Q_1;\ d_1 + c < y^*_t(t,x) < d_2 - c\}$.

For sufficiently small c, mes $Q_1^c > 0$. Since $y^\varepsilon_t \to y^*_t$ strongly in $C(0,T;L^2(\Omega))$, the Egorov theorem shows that, for any $\eta > 0$ sufficiently small, there is $Q_\eta \subset Q_1^c$ (c fixed) such that $\operatorname{mes}(Q_1^c - Q_\eta) < \eta$ and $y^\varepsilon_t \to y^*_t$ uniformly on Q_η.

For $\varepsilon > 0$ sufficiently small, it yields that $d_1 + c/2 \le y^\varepsilon_t(t,x) \le d_2 - c/2$ on Q_η and $\dot{\beta}^\varepsilon_1(y^\varepsilon_t(t,x)) = 0$ a.e. Q_η. This is a consequence of the approximation properties of β^ε_1 and of the fact that β_1 is constant on the interval (d_1, d_2). Obviously $Q_1^c = \bigcup_{\eta>0} Q_\eta$ and by the above argument we get (2.7).

Consider now Ω to be a subdomain of R. We know that y^ε_t, $y^*_t \in L^\infty(0,T;H_0^1(\Omega)) \cap W^{1,2}(0,T;L^2(\Omega))$. Then y^ε_t, y^*_t are continuous on Q and Q_0 is an open set. Moreover $y^\varepsilon_t \to y^*_t$ uniformly on $\bar{Q}$. This may be infered from Lions' lemma, Ch.I, § 1, as follows:

As $H^1(\Omega) \subset C(\bar{\Omega})$ we have that for any $\mu > 0$ there is $C(\mu) > 0$:

$$\sup_{x \in \bar{\Omega}} |y^\varepsilon_t(t,x) - y^*_t(t,x)| \le \mu |y^\varepsilon_t(t,.) - y^*_t(t,.)|_{H^1(\Omega)} + C(\mu) |y^\varepsilon_t(t,.) - y^*_t(t,.)|_{L^2(\Omega)}.$$

We take the sup over $t \in [0,T]$:

(2.8) $$|y^\varepsilon_t - y^*_t|_{C(\bar{Q})} \le \mu |y^\varepsilon_t - y^*_t|_{L^\infty(0,T;H^1(\Omega))} + C(\mu) |y^\varepsilon_t - y^*_t|_{C(0,T;L^2(\Omega))}.$$

Since $\{y_t^\varepsilon\}$ is bounded in $L^\infty(0,T;H_0^1(\Omega))$ and $y_t^\varepsilon \to y_t^*$ strongly in $C(0,T;L^2(\Omega))$, by (2.8) we see that $y_t^\varepsilon \to y_t^*$ uniformly on $\bar{Q}$.

Assume now that $K \subset Q_1$ is compact (Q_1 is open here). There is $c > 0$ such that $d_1 + c \leq y_t^* \leq d_2 - c$ on K and, as above, for ε sufficiently small, we get $\dot{\beta}_1^\varepsilon(y_t^\varepsilon) = 0$ on K. This shows that δ_1 vanishes on Q_0.

As concerns the fact that $\dot{\beta}_2^\varepsilon(y_t^\varepsilon)m_t^\varepsilon \to D\beta_2(y_t^*)m_t^*$ this follows in a standard way by means of the use of a saddle function, according to Chapter II, § 3.

Remark 2.2. The condition that $\Omega \subset R$ is essential here because there are no $L^p(\Omega)$ regularity results for hyperbolic variational inequalities.

Remark 2.3. Let $L(y,u) = 1/2\,|y|^2_{L^2(\Omega)} + I_{U_{ad}}$, where $U_{ad} = \{u \in L^2(\Omega);\ 0 \leq u(x) \leq 1$ a.e.$\Omega\}$. Then (2.6) may be equivalently written as

$$-\int_Q B^* m_t^*(v - u^*) \leq 0 \qquad v \in \tilde{U}_{ad},$$

with $\tilde{U}_{ad} = \{u \in L^2(Q); u(t,.) \in U_{ad}$ a.e.$[0,T]\}$.

Obviously $u^*(t,x) = 0$ a.e. in the set where $B^* m_t^* > 0$ and $u^*(t,x) = 1$ a.e. in the set where $B^* m_t^* < 0$. However, on the set where $B^* m_t^* = 0$, which generally may be very rich, we cannot obtain information on u^* since the regularity of m_t^* is very weak and the reasoning from Corollary 1.9 doesn't apply.

Remark 2.4. Again from (2.9), as in Remark 3.17 from Chapter II, it yields that in the absence of control constraints and for quadratic cost functional L, all the optimal controls u^* belong to $L^\infty(0,T;U)$.

Remark 2.5. A special case of Thm.2.1 was discussed in the paper [133].

3. The vibrating string with obstacle

We consider the equation of the infinite string in the form

(3.1) $$y_{tt} - y_{xx} + \beta(x,y) \ni u \qquad \text{in }]0,T[\times R,$$

(3.2) $$y(0,x) = y_0(x),\ \ y_t(0,x) = v_0(x) \qquad \text{in } R,$$

where $\beta(x,.)$ is the maximal monotone graph

(3.3) $$\beta(x,y) = \begin{cases} 0 & y > \psi(x), \\ ]-\infty,0] & y = \psi(x), \\ \emptyset & y < \psi(x) \end{cases}$$

and ψ is a continuous function on R.

Therefore the string is forced to vibrate above the obstacle $y = \psi(x)$. The physical significance of the nonlinear term $\beta(x,y(t,x))$ is the unknown reaction of the obstacle and we formulate the problem to minimize this reaction with minimal cost u. So, we study the optimal control problem

$$(3.4)\qquad \text{Minimize}\left\{|\beta(y)|_{L^2(\Pi_T)} + |u|_{L^2(\Pi_T)}\right\}$$

subject to $u,y\in L^2(\Pi_T)$ satisfying (3.1), (3.2).

Here $\Pi_T = [0,T]\times R$, $v_0\in L^2_{loc}(R)$, $y_0\in H^1_{loc}(R)$ and $y_0(x)\geq \varphi(x)$.

Starting with the work of Amerio and Prouse [7], the equation (3.1), (3.2) was studied by Amerio [6], C.Citrini [33], M.Schatzman [104]. The main tool is the notion of line of influence of the obstacle and a solution is built step by step, following the characteristics. This method is difficult to use here and we adopt the point of view from the theory of unstable control systems as in section 4, Chapter II. We also underline that it is not possible to apply the abstract scheme from section 1, Chapter II.

The control $u\in L^2(\Pi_T)$ is called admissible if there are $y\in L^2(0,T;H^1_{loc}(R))$, $w\in L^2(\Pi_T)$, $w\in\beta(y)$ a.e., such that $y(0,x) = y_0(x)$ a.e. and:

$$(3.5)\qquad \int_{\Pi_T}(\operatorname{grad} y \operatorname{grad} v + w.v - y_t.v_t)dxdt = \int_{\Pi_T} u.v.dxdt + \int_R v_0(x).v(x,0)dx$$

for any $v\in H^1(\Pi_T)$ with compact support and with $v(T,x) = 0$, $x\in R$.

The pair $[y,w]$ is called a generalized solution of (3.1), (3.2). The following inequality, called the "energy estimate", V.Mikhailov [77], p.332, will be of frequent use in this section:

$$(3.6)\qquad \Big[\int_{|x|<r-t}(z_t^2 + |\operatorname{grad} z|^2)(t,x)dx\Big]^{1/2} \leq \Big[\int_{|x|<r}(z_t^2+|\operatorname{grad} z|^2)(0,x)dx\Big]^{1/2} +$$

$$+ 2\sqrt{t}\,|f|_{L^2\{(x,\tau);\ |x|<r-\tau,\ 0<\tau<T\}},$$

where z is the generalized solution of the linear equation $z_{tt} - z_{xx} = f$, with the condition (3.2).

The notion of solution used in [6], [7], [104] allows w to be a distribution with support in the contact set $\{(t,x)\in\Pi_T;\ y(t,x) = \varphi(x)\}$. This is written here in the form $w\in\beta(y)$ and the condition $w\in L^2(\Pi_T)$ is in fact a constraint on the set of admissible pairs. We remark that the set of admissible pairs is sufficiently rich and the problem (3.4) is nontrivial.

This may be seen as follows. If $[y,u]$ is an admissible pair and $[y,w]$ is the generalized solution correspoding to u, then $[y,u-w]$ is again an admissible pair with $[y,0]$ being the generalized solution corresponding to u-w. The d'Alembert formula shows that any control v from $L^2(\Pi_T)$ which majorizes a.e. u-w is admissible since the associated solution z of the wave equation stisfies $z\geq y$ a.e. Π_T, that is $z\geq\varphi$ a.e. Π_T and the pair $[z,0]$ is then a generalized solution of (3.1), (3.2) corresponding to v.

One can easily obtain examples of admissible controls if φ is regular and the initial conditions are well choosen.

Proposition 3.1. The existence of an admissible control implies the existence of at least one optimal pair $[y^*,u^*]$ for the problem (3.1) - (3.4).

Proof

We denote by J(u) the cost functional (3.4). Then inf J(u) is finite and positive. Let $\{u_n\}$ be a minimizing sequence in $L^2(\Pi_T)$. By (3.4), $\{u_n\}$ and $\{w_n\}$ are bounded in $L^2(\Pi_T)$, where $w_n \in \beta(y^n)$ and $[y^n, w_n]$ is the generalized solution of (3.1), (3.2) corresponding to u_n.

The energy estimate (3.6) shows that $\{y^n\}$, $\{y_t^n\}$ are bounded in $L^\infty(0,T;H^1_{loc}(R))$, $L^\infty(0,T;L^2_{loc}(R))$ respectively. So, on a subsequence depending upon K, we have

$$y^n \to y \text{ strongly in } L^2(\Omega_K \times [0,T]),$$

where Ω_K is a nondecreasing sequence of compact subsets and $U\Omega_K = R$. The limit y is independent of K because, passing from K to K+1 we take further subsequences, if necessary.

On the other hand $\beta(y^n) \to w$ weakly in $L^2(\Pi_T)$ and $w(t,x) \in \beta(y(t,x))$ a.e. Π_T since the operator obtained from (3.3) in $L^2(\Omega_K \times [0,T])$ is maximal monotone, according to Brezis [18], p.25. We also have $u_n \to u$ weakly in $L^2(\Pi_T)$ and

$$\text{(3.7)} \qquad \liminf J(u_n) \geq |w|_{L^2(\Pi_T)} + |u|_{L^2(\Pi_T)}.$$

It is easy to pass to the limit on a subsequence and to see that [y,w] is the generalized solution of (3.1), (3.2) corresponding to u. From (3.7) it yields that [y,u] is an optimal pair, which we denote $[y^*, u^*]$.

We define the approximate problem

$$\text{(3.8)} \qquad \text{Minimize } \left\{ |\beta^\varepsilon(y)|_{L^2(\Pi_T)} + |u|_{L^2(\Pi_T)} \right\},$$

$$\text{(3.9)} \qquad y_{tt} - y_{xx} + \beta^\varepsilon(x,y) = u$$

and (3.2). Here β^ε is a regularization of β given by

$$\text{(3.10)} \qquad \beta^\varepsilon(x,y) = \int_{-\infty}^{\infty} \beta_\varepsilon(x, y + \varepsilon^2 - \varepsilon^2\tau)\rho(\tau)d\tau, \quad \varepsilon > 0,$$

and β_ε is the Yosida approximation of β. As usual ρ is a Friedrichs mollifier, i.e. $\rho \geq 0$, $\rho(-\tau) = \rho(\tau)$, $\text{supp}\rho \subset [-1,1]$, $\rho \in C^\infty(R)$ and $\int_{-\infty}^{\infty} \rho(\tau)d\tau = 1$.

We remark that, by (3.10), we have $\beta^\varepsilon(x,y) = 0$ for $y \geq \varphi(x)$, $y \in R$.

Proposition 3.2. If $y_0 \in H^1_{loc}(R)$, $v_0 \in L^2_{loc}(R)$ and $u \in L^2(0,T;L^2_{loc}(R))$, the equation (3.9), (3.2) has a unique generalized solution $y \in L^\infty(0,T;H^1_{loc}(R))$, $y_t \in L^\infty(0,T;L^2_{loc}(R))$.

Proof (sketch)

This is quite a standard result and we only sketch the proof. Let $y_0^n \in H^1(R)$, $v_0^n \in L^2(R)$ and $u^n \in L^2(\Pi_T)$ be some approximations of the data (obtained for instance by truncation) and $y^n \in L^\infty(0,T;H^1(R)) \cap W^{1,\infty}(0,T;L^2(R))$ be the generalized solution of (3.9) corresponding to u_n. The existence and the uniqueness of y^n follow from the d'Alembert formula and a fixed point argument. The regularity is a consequence of the results for linear hyperbolic equations.

By the energy estimate and the Gronwall inequality we obtain $\{y^n\}$ bounded in $L^\infty(0,T;H^1_{loc}(R))\cap W^{1,\infty}(0,T;L^2_{loc}(R))$ since β^ε is Lipschitzian. It also yields that $\{\beta^\varepsilon(y^n)\}$ is bounded in $L^\infty(0,T;L^2_{loc}(R))$. To pass to the limit, the same argument as in Proposition 3.1 may be used. The uniqueness is obvious.

It is possible to prove the existence of approximate optimal pairs $[y^\varepsilon,u_\varepsilon]$ along the same lines as in Proposition 3.1.

Proposition 3.3. We have

(i) $J_\varepsilon(u_\varepsilon) \leq J(u^*)$

(ii) $\lim\limits_{\varepsilon\to 0} J_\varepsilon(u_\varepsilon) = J(u^*)$

(iii) on a subsequence

$u_\varepsilon \to u^*$ strongly in $L^2(\ _T)$

$\beta^\varepsilon(y^\varepsilon)\to w^*\in\beta(y^*)$ strongly in $L^2(\Pi_T)$,

$y^\varepsilon \to y^*$ strongly in $C(0,T;L^2_{loc}(R))$,

where $[y^*, u^*]$ is an optimal pair for the problem (3.1) - (3.4) and J, J_ε denote the cost functionals of the problems (3.1) - (3.4), respectively (3.8), (3.9).

Proof.

The pair $[y^*,u^* - w^*]$ is admissible for the approximate problem since $y^*(t,x) \geq \varphi(x)$ a.e. Π_T, so $\beta^\varepsilon(x, y^*(t,x)) = 0$ a.e. Π_T. Then

(3.11) $$J_\varepsilon(u^* - w^*) = |u^* - w^*|_{L^2(\Pi_T)} \leq J(u^*)$$

and (i) follows.

By (3.11) and the definition of J_ε we see that $\{u_\varepsilon\}$, $\{\beta^\varepsilon(y^\varepsilon)\}$ are bounded in $L^2(\Pi_T)$. Then (3.9) and (3.6) imply that $\{y^\varepsilon\}$ is bounded in $L^\infty(0,T;H^1_{loc}(R))$ and $\{y^\varepsilon_t\}$ is bounded in $L^\infty(0,T;L^2_{loc}(R))$. On a subsequence, again denoted ε, we have

(3.12) $$\begin{aligned} &u_\varepsilon \to u && \text{weakly in } L^2(\Pi_T),\\ &\beta^\varepsilon(y^\varepsilon)\to w && \text{weakly in } L^2(\Pi_T),\\ &\liminf_{\varepsilon\to 0} J_\varepsilon(u_\varepsilon) \geq |u|_{L^2(\Pi_T)} + |w|_{L^2(\Pi_T)}. \end{aligned}$$

Let y be the strong limit in $C(0,T;L^2_{loc}(R))$ of y^ε. The same argument as in Proposition 3.1 shows that $w\in\beta(y)$ a.e. Π_T and that the pair $[y,u]$ is admissible for the problem (3.1) - (3.4). Combining (3.12) and (i), it yields that $[y,u]$ is an optimal pair and (ii) is proved. Since the optimal value $J(u^*)$ is unique, the convergence is valid without taking subsequences.

The last part is a consequence of a wellknown strong convergence criterion in uniformly convex Banach spaces.

Let us suppose now that $y_0\in H^1(R))$ and $v_0\in L^2(R)$. Then $y^\varepsilon\in L^\infty(0,T;H^1(R))$ $\cap W^{1,\infty}(0,T;L^2(R))$ and any $u\in L^2(\Pi_T)$ is admissible for the problem (3.8), (3.9).

We denote by $\Psi: L^2(\Pi_T)\to R$ the convex, continuous functional, $\Psi(u) = |u|_{L^2(\Pi_T)}$.

Theorem 3.4. If $\varphi \in C(R)$, there is an optimal pair $[y^*,u^*] \in L^2(0,T;H^1(R)) \times L^2(\Pi_T)$, an optimal adjoint state $p^* \in L^2(\Pi_T)$ and a distribution δ on Π_T with $\operatorname{supp}\delta \subset \{(t,x) \in \Pi_T;\ y^*(t,x) = \varphi(x)\}$ satisfying the optimality system:

(3.13) $\quad p^*_{tt} - p^*_{xx} = \delta \quad$ in Π_T,

(3.14) $\quad p^* \in \partial\Psi(u^*) \quad$ in Π_T.

For the proof we need the following proposition which is a particular case of the results from Chapter II, §4.

Proposition 3.5. For any solution $[y^\varepsilon, u_\varepsilon]$ of the problem (3.8), (3.9) there is $p^\varepsilon \in L^\infty(0,T;H^1(R)) \cap W^{1,\infty}(0,T;L^2(R))$ such that

(3.15) $\quad p^\varepsilon_{tt} - p^\varepsilon_{xx} + \dot\beta^\varepsilon(y^\varepsilon)p^\varepsilon \in -\partial\Psi(\beta^\varepsilon(y^\varepsilon))\dot\beta^\varepsilon(y^\varepsilon) \quad$ in Π_T,

(3.16) $\quad p^\varepsilon(T,x) = 0,\ p^\varepsilon_t(T,x) = 0 \quad$ in R,

(3.17) $\quad p^\varepsilon \in \partial\Psi(u_\varepsilon) \quad$ in Π_T.

Proof of Thm.3.4

By (3.17), Proposition 3.3 (iii) and the definition of Ψ, we see that $\{p^\varepsilon\}$ is bounded in $L^2(\Pi_T)$ and, on a subsequence, $p^\varepsilon \to p^*$ weakly in $L^2(\Pi_T)$. Taking further subsequences we know that $u_\varepsilon \to u^*$ strongly in $L^2(\Pi_T)$ and $y^\varepsilon \to y^*$ strongly in $C(0,T;L^2_{loc}(R))$.

The relation (3.14) is an immediate consequence of (3.17) and the demiclosedness of $\partial\Psi$. As concerns (3.13) we remark that $\{y^\varepsilon\}$ is bounded in $L^\infty(0,T;H^1(R)) \cap W^{1,\infty}(0,T;L^2(R))$.

This implies that y^ε, y^* are continuous in Π_T and $y^\varepsilon \to y^*$ uniformly on compact subsets of Π_T, as in §2.

Consider $Q^n_\mu = \{(t,x) \in \Pi_T;\ -n < x < n,\ y^*(t,x) > \varphi(x) + \mu\}$ and $Q_0 = \{(t,x) \in \Pi_T;\ y^*(t,x) > \varphi(x)\}$, open subsets of Π_T. There is $\varepsilon_0 > 0$, such that, for $\varepsilon < \varepsilon_0$, $y^\varepsilon(t,x) \geq \varphi(x) + \mu/2$ on Q^n_μ, so $p^\varepsilon_{tt} - p^\varepsilon_{xx} = 0$ on Q^n_μ.

Passing to the limit in $\mathcal{D}'(\Pi_T)$ we see that the distribution $p^*_{tt} - p^*_{xx}$ vanishes on Q^n_μ. Since $Q_0 = \bigcup_{n,\mu} Q^n_\mu$, the proof is finished.

Remark 3.6. The results of this section closely follow the author's papers [120], [121]. Partial extensions to higher space dimensions or to bounded domains (with Dirichlet boundary conditions for instance) are possible. More general cost functionals including terms of the form $|y - y_d|_{L^2(\Pi_T)}$ or $|y(T) - y_d|_{L^2(R)}$ may be also considered.

4. The variational inequality method

As we have mentioned at the beginning of this chapter, there is a strong relationship between the state constrained control problems and the problems governed by variational inequalities. In some special cases, the two types of problems are equivalent.

This gives a new interpretation and motivation for the control problems governed by variational inequalities and a new efficient approximation method for the constrained control problems.

To clarify these ideas, we study the following simple model problem:

$$\text{(4.1)} \qquad \text{Minimize}\{g(y) + h(u)\},$$

$$\text{(4.2)} \qquad y' + Ay = Bu + f \qquad \text{in }]0,T[,$$

$$\text{(4.3)} \qquad y(0) = y_0,$$

$$\text{(4.4)} \qquad y(t) \in C \qquad \text{in } [0,T].$$

Here, we take V,H,U to be Hilbert spaces with $V \subset H \subset V^*$ densely and compactly, and $A : V \to V^*$, $B : U \to H$ are linear, bounded operators, such that:

$$\text{(4.5)} \qquad (Au,u) \geq \omega |u|_V^2, \quad \omega > 0, \quad u \in V,$$

$$\text{(4.6)} \qquad (Au,v) = (u,Av), \quad u,v \in V.$$

We denote by (.,.) the pairing between V and V^* (if v_1, $v_2 \in H$, then (v_1,v_2) is the inner product in H) and we assume that $C \subset H$ is a closed, convex subset, $y_0 \in C$, $Ay_0 \in H$, $f \in L^2(0,T;H)$. The mappings $g : L^2(0,T;H) \to R$, $h : L^2(0,T;U) \to]-\infty,+\infty]$ are convex, proper.

Moreover g is continuous, majorized from below by a constant, while h is lower semicontinuous and coercive:

$$\text{(4.7)} \qquad \lim_{|u|_{L^2(0,T;U)} \to \infty} h(u) = +\infty.$$

Under these assumptions, the equation (4.2),(4.3) has a unique solution $y \in C(0,T;V)$, $y' \in L^2(0,T;H)$ and (4.4) makes sense.

If control constraints, $u \in U_0$ (a convex, closed subset of $L^2(0,T;U)$) are imposed, they may be implicitly expressed, as usual, by adding to h the indicator function of U_0 in $L^2(0,T;U)$. This modification preserves the hypotheses on h.

We suppose the existence of an admissible pair $[\bar{y}, \bar{u}]$ for the problem (4.1). This condition may be relaxed, according to (4.29).

It is standard to prove the existence of at least one optimal pair $[y^*, u^*]$ by the coercivity assumption (4.7).

Let $\varphi : H \to]-\infty,+\infty]$ be the convex, lower semicontinous, proper function, given by

$$\text{(4.8)} \qquad \varphi(y) = \begin{cases} 0 & y \in C, \\ +\infty & \text{otherwise.} \end{cases}$$

The variational inequality method (a variant [130]) consists in associating to the problem (4.1), the approximating problem

$$\text{(4.9)} \qquad \text{Minimize}\left\{g(y) + h(u) + 1/2\, |w|^2_{L^2(0,T;V^*)}\right\},$$

(4.10) $y' + Ay + \varepsilon w = Bu + f, \quad w \in \partial\tilde{\varphi}(y), \ \varepsilon > 0,$

and (4.3). Here $\tilde{\varphi}: V \to]-\infty,+\infty]$ is given by $\tilde{\varphi}(v) = \varphi(v)$.

The equation (4.10) is a variational inequality and has a unique solution $y \in C(0,T;H) \cap L^\infty(0,T;V)$, $y' \in L^2(0,T;H)$, for any $u \in L^2(0,T;U)$, according to Barbu [12], p.189.

Every $u \in L^2(0,T;U)$ (satisfying the control constraints, if any) is admissible for the problem (4.4), (4.10). Furthermore, denoting by $J_\varepsilon(y,u)$ the functional (4.9) and by $J(y,u)$ the functional (4.1) we have

$$J_\varepsilon(\bar{y},\bar{u}) = J(\bar{y},\bar{u})$$

for all the admissible pairs $[\bar{y},\bar{u}]$ of (4.2) - (4.4) since $\bar{w} \in \partial\tilde{\varphi}(\bar{y})$ may be taken zero by $\bar{y}(t) \in C$.

Let us denote by $[y_\varepsilon, u_\varepsilon]$ an optimal pair for the problem (4.9), (4.10), which obviously exists under condition (4.7).

<u>Proposition 4.1. On a subsequence</u> $\varepsilon \to 0$ <u>we have:</u>

(4.11) $u_\varepsilon \to u^*$ <u>weakly in</u> $L^2(0,T;U)$,

(4.12) $y_\varepsilon \to y^*$ <u>strongly in</u> $C(0,T;H)$,

(4.13) $J_\varepsilon(y_\varepsilon, u_\varepsilon) \to J(y^*,u^*)$

<u>where</u> $[y^*,u^*]$ <u>is an optimal pair for</u> (4.1).

<u>Proof</u>

As $[y^*,u^*]$ is admissible for (4.1) then $0 \in \partial\tilde{\varphi}(y^*)$ and we have

$$J(y^*,u^*) = J_\varepsilon(y^*,u^*) \geq J_\varepsilon(y_\varepsilon, u_\varepsilon),$$

and the assumptions on h,g imply that $\{u_\varepsilon\}$ is bounded in $L^2(0,T;U)$, $\{w_\varepsilon\}$ is bounded in $L^2(0,T;V^*)$. By (4.9) we obtain $\{y_\varepsilon\}$ bounded in $L^\infty(0,T;V)$, $\{y'_\varepsilon\}$ bounded in $L^2(0,T:H)$. Let $[\tilde{y},\tilde{u}]$ be the weak limit in $L^2(0,T;V) \times L^2(0,T;U)$ of $[y_\varepsilon, u_\varepsilon]$. We have $y_\varepsilon \to \tilde{y}$ strongly in $C(0,T;H)$, $g(y_\varepsilon) \to g(\tilde{y})$ since g is continuous and

$$\liminf_{\varepsilon \to 0} h(u_\varepsilon) \geq h(\tilde{u}),$$

(4.14) $g(\tilde{y}) + h(\tilde{u}) \leq J(y^*,u^*).$

Since $\varepsilon w_\varepsilon \to 0$ strongly in $L^2(0,T;V^*)$, we see that $\tilde{y}$ is the solution of (4.2), (4.3) corresponding to $\tilde{u}$. Moreover $\tilde{y}(t) \in C$ because $y_\varepsilon(t) \in C \supseteq \mathrm{dom}(\partial\tilde{\varphi})$ and $y_\varepsilon \to \tilde{y}$ strongly in $C(0,T;H)$.

By (4.14) it yields that $[\tilde{y},\tilde{u}]$ is an optimal pair for the problem (4.1), which we denote $[y^*,u^*]$. This proves (4.11), (4.12). As concerns (4.13), we have

$g(\tilde{y}) + h(\tilde{u}) \leq \liminf J_\varepsilon(y_\varepsilon, u_\varepsilon) \leq \limsup J_\varepsilon(y_\varepsilon, u_\varepsilon) \leq J(y^*, u^*)$.

Remark 4.2. Let $\tilde{w}$ be the weak limit in $L^2(0,T;V^*)$ of w_ε. By (4.13) and the continuity of g, we have $h(u_\varepsilon) \to h(u^*)$, so $\tilde{w} = 0$ and $w_\varepsilon \to 0$ strongly in $L^2(0,T;V^*)$.

Remark 4.3. If h is uniformly convex, we get $u_\varepsilon \to u^*$ strongly in $L^2(0,T;U)$, on a subsequence. In function spaces, it is enough for h to be strictly convex and to satisfy a growth condition, stronger than (4.7), according to Visintin [136].

Let y^ε be the solution of (4.2), (4.3) corresponding to u_ε. The pair $[y^\varepsilon, u_\varepsilon]$ isn't necessarily admissible for the problem (4.1) - (4.4), but we may compute $J(y^\varepsilon, u_\varepsilon)$ and prove the following suboptimality result:

Corollary 4.4. We have

$$\lim_{\varepsilon \to 0} J(y^\varepsilon, u_\varepsilon) = J(y^*, u^*), \tag{4.15}$$

$$\mathrm{dist}(y^\varepsilon, C \cap V)_{L^2(0,T;V) \cap C(0,T;H)} \leq K.\varepsilon, \tag{4.16}$$

where K is independent of $\varepsilon > 0$.

Proof

We denote $z_\varepsilon = y^\varepsilon - y_\varepsilon$ and we obtain

$$z'_\varepsilon + Az_\varepsilon = \varepsilon w_\varepsilon \qquad \text{in } [0,T], \tag{4.17}$$
$$z_\varepsilon(0) = 0.$$

Multiply (4.17) by z_ε and integrate over [0,T]. By the properties of A, w_ε, we infer

$$|z_\varepsilon|_{C(0,T;H) \cap L^2(0,T;V)} \leq K\varepsilon.$$

But $y_\varepsilon(t) \in C \cap V = \mathrm{dom}(\partial\tilde{\varphi})$ and (4.16) follows. Moreover, it yields that $y^\varepsilon - y_\varepsilon \to 0$ in $C(0,T;H)$. By Remark 4.2 and the continuity of g, we deduce (4.15) too.

Remark 4.5. It may be difficult to find u_ε by a usual gradient algorithm since the problem (4.9), (4.10) is not a differentiable optimization problem. To overcome this difficulty we use a smoothing procedure.

Let $\beta^\lambda : H \to H$ satisfy

$$\beta^\lambda(y) = 0 \qquad \text{for } y \in C, \tag{4.18}$$

$$|\beta^\lambda(y) - (\partial\varphi)_\lambda(y)|_H \leq c\lambda, \quad \lambda > 0, \tag{4.19}$$

$$\beta^\lambda \text{ is Gateaux differentiable,} \tag{4.20}$$

where $(\partial\varphi)_\lambda$ is the Yosida approximation of the operator $\partial\varphi : H \to H$. See for instance (3.10) for an example of such a regularization in $L^2(R)$. Other examples may be similarly obtained starting, for instance, from the Examples 1-3, § 1.

We consider the regularized problem

(4.21) $$\text{Minimize}\left\{g(y) + h(u) + 1/2\,|\beta^{\lambda}(y)|^2_{L^2(0,T;H)}\right\},$$

(4.22) $$y' + Ay + \varepsilon\beta^{\lambda}(y) = Bu + f$$

and (4.3).

We denote $[y_\lambda, u_\lambda]$ an optimal pair of (4.21), (4.22) and let y^λ be the solution of (4.2), (4.3) corresponding to u_λ.

Corollary 4.6. We have

(4.23) $$\text{dist}(y^\lambda(t), C)_H \le K\ + c\sqrt{\lambda/\varepsilon}\,,$$

(4.24) $$J(y^\lambda, u_\lambda) \le J(y^*, u^*) + \eta_\lambda(\varepsilon),$$

where $\eta_\lambda(\varepsilon) \to 0$ for $\varepsilon \to 0$ and K, c are constants independent of $\lambda, \varepsilon > 0$.

Proof

Since $y^*(t) \in C$, then $\beta^\lambda(y^*(t)) = 0$ for $t \in [0,T]$ and $[y^*, u^*]$ is an admissible pair for (4.21), (4.22) with $J_{\varepsilon,\lambda}(y^*, u^*) = J(y^*, u^*)$. It yields that $\{u_\lambda\}$ is bounded in $L^2(0,T;U)$ and $\{\beta^\lambda(y_\lambda)\}$ is bounded in $L^2(0,T;H)$. We multiply (4.22) by y'_λ and we integrate over $[0,t]$:

$$\int_0^t |y'_\lambda|^2 d\tau + \int_0^t (Ay_\lambda, y'_\lambda) d\ \ + \varepsilon\varphi_\lambda(y_\lambda(t)) \le \int_0^t (Bu_\lambda + f, y'_\lambda) + c.\lambda$$

from (4.19) and the chain rule.

Then $\{y_\lambda\}$ is bounded in $L^\infty(0,T;V)$, $\{y'_\lambda\}$ is bounded in $L^2(0,T;H)$, $\{\varepsilon\varphi_\lambda(y_\lambda)\}$ is bounded in $L^\infty(0,T)$ by constants independent of $\lambda, \varepsilon > 0$.

Let $z_\lambda = y^\lambda - y_\lambda$. An argument similar to (4.17) shows that $|z_\lambda(t)|_H \le K\varepsilon$. We also remark that

$$\varphi_\lambda(y) = 1/2\lambda\, \text{dist}(y, C)^2_H.$$

Combining these facts with the boundedness of $\{\varepsilon\varphi_\lambda(y_\lambda)\}$ in $L^\infty(0,T)$, we obtain (4.23). As concerns (4.24), we have

$$J(y^\lambda, u_\lambda) \le J_{\varepsilon,\lambda}(y_\lambda, u_\lambda) + g(y^\lambda) - g(y_\lambda) \le J(y^*, u^*) + g(y^\lambda) - g(y_\lambda)$$

By the estimate for z_λ, $\{y_\lambda\}$ and $\{y^\lambda\}$ are relatively compact subsets in $C(0,T;H)$ and converge to the same limit. As g is locally Lipschitzian on $L^2(0,T;H)$ we obtain $|g(y^\lambda) - g(y_\lambda)| \le \eta_\lambda(\varepsilon)$, $\lambda, \varepsilon > 0$ and the proof is finished.

Remark 4.7. If g is Lipschitzian on bounded subsets of $L^2(0,T;H)$ (situation frequently met in the applications), then $\eta_\lambda(.)$ is independent of λ.

Remark 4.8. It is possible to take $|w|^2_{L^2(0,T;H)}$ directly in (4.9). Then (4.9), (4.10) has to be interpreted as a singular control problem, according to § 4, Chapter II.

Remark 4.9. By Corollary 4.6 we see that u_λ gives a suboptimal solution for the problem (4.1). To compute it a gradient algorithm may be used.

Let us briefly compare this approach with the usual penalization method, when applied to the problem (4.1).The approximate problem to be considered is

$$\text{(4.25)} \qquad \text{Minimize}\left\{g(y) + h(u) + \int_0^T \varphi_\lambda(y)dt\right\},$$

subject to (4.2), (4.3) and with φ_λ as above. We denote $[\tilde{y}^\lambda, u^\lambda]$ an optimal pair of (4.25). The following result may be easily obtained (Barbu - Precupanu [14], Ch.IV):

Proposition 4.10. On a subsequence, we have:

i) $u^\lambda \to \hat{u}$ weakly in $L^2(0,T;U)$,

ii) $\tilde{y}^\lambda \to \hat{y}$ strongly in $C(0,T;H)$,

iii) $\text{dist}(\tilde{y}^\lambda, C)_{L^2(0,T;H)} \le c\,\lambda^{1/2}$,

where $[\hat{y},\hat{u}]$ is an optimal pair for (4.1) - (4.4) and c is independent of $\lambda > 0$.

Let us fix $\varepsilon = \lambda^{1/2}$ in (4.23). We see that the variational inequality approach produces pointwise estimates of the violation of the constraints, while the penalization method gives only integral estimates.

However, we remark that (4.23), (4.16) give pointwise estimates with respect to $t \in [0,T]$ in the H norm, which is an integral norm, generally.

In the setting of partial differential equations, the variational inequality approach gives pointwise estimates also with respect to the space variables. We argue this by means of the following parabolic constrained control problem:

$$\text{(4.26)} \qquad \text{Minimize}\left\{ g(y) + h(u)\right\}$$

$$\text{(4.27)} \qquad y_t - \Delta y = Bu + f \qquad \text{in } Q =]0,T[\times \Omega,$$

$$\text{(4.28)} \qquad y(t,x) = 0 \qquad \text{in } \Sigma = [0,T] \times \partial\Omega,$$

$$\text{(4.29)} \qquad y(0,x) = y_0(x) \qquad \text{in } \Omega,$$

$$\text{(4.30)} \qquad u \in U_{ad},$$

$$\text{(4.31)} \qquad y(t,x) \in [a,b] \qquad \text{in } Q.$$

We have denoted as usual by Ω a bounded domain in R^N with smooth boundary $\partial\Omega$ and we assume that $f \in L^\infty(Q)$, $0 \in [a,b]$, $y_0(x) \in [a,b]$ a.e.Ω, $U_{ad} \subset U$ is a closed convex set such that BU_{ad} is a bounded subset of $L^\infty(Q)$. The assumptions on g, h, B are as at the beginning of this section, with $H = L^2(\Omega)$.

Under a standard admissibility condition, one may infer the existence of at least one optimal pair $[y^*,u^*]$ in $C(0,T;H) \times L^2(0,T;U)$.

Let β be the subdifferential of the indicator function of $[a,b]$ in $R \times R$ and $p > \frac{N+2}{2}$ be given.

The approximating problem has the following form

$$\text{(4.32)} \qquad \text{Minimize} \left\{ g(y) + h(u) + \frac{1}{\varepsilon} |\beta_\varepsilon(y)|_{L^p(Q)} \right\}$$

over all the pairs $[y,u]$ satisfying $u \in L^2(0,T;U)$,

$$\text{(4.33)} \qquad y_t - \Delta y + \beta_\varepsilon(y) = Bu + f \qquad \text{in } Q$$

and (4.28)-(4.30).

We notice the similarity with the problem (4.21), (4.22). Denote $[y_\varepsilon, u_\varepsilon]$ an optimal pair for the problem (4.32). Obviously $\{Bu_\varepsilon\}$ is bounded in $L^\infty(Q)$ since $u_\varepsilon \in U_{ad}$, $\varepsilon > 0$.

Any admissible pair for (4.26) is again admissible for (4.32) with the same cost because the nonlinear penalization term $\beta_\varepsilon(\cdot)$ vanishes on admissible states. Then, (4.32) implies that

$$\text{(4.34)} \qquad |\beta_\varepsilon(y_\varepsilon)|_{L^p(Q)} \leq C\varepsilon .$$

Let y^ε be the solution of (4.27)-(4.29) corresponding to u_ε and $z^\varepsilon = y^\varepsilon - y_\varepsilon$. We have

$$\begin{aligned} z_t^\varepsilon - \Delta z^\varepsilon &= \beta_\varepsilon(y_\varepsilon) && \text{in } Q, \\ z^\varepsilon(t,x) &= 0 && \text{in } \Sigma, \\ z^\varepsilon(0,x) &= 0 && \text{in } \Omega, \end{aligned}$$

and, by (4.34), we infer

$$|z^\varepsilon|_{W^{2,1,p}(Q)} \leq C\varepsilon ,$$

that is

$$\text{(4.35)} \qquad |y^\varepsilon(t,x) - y_\varepsilon(t,x)| \leq C\varepsilon , \qquad (t,x) \in Q,$$

by the condition $p > \frac{N+2}{2}$ and the imbedding theorem (Ladyzhenskaya, Uralceva, Solonnikov [67]).

Now, we use in (4.33) Thm. 4.5., Ch. I, and we obtain that

$$|\beta_\varepsilon(y_\varepsilon)|_{L^q(Q)} \leq |Bu \;\; + f|_{L^q(Q)}$$

for any $q > 1$. Taking $q \to \infty$, as $\{Bu_\varepsilon + f\}$ is bounded in $L^\infty(Q)$, we get that $\{\beta_\varepsilon(y_\varepsilon)\}$ is bounded in $L^\infty(Q)$.

We recall that

$$\beta_\varepsilon(y) = \frac{y - \mathrm{Proj}_{[a,b]}(y)}{\varepsilon},$$

therefore, we have that

(4.36) $\qquad \mathrm{dist}(y_\varepsilon(t,x),[a,b]) \le C\varepsilon, \quad (t,x) \in Q.$

Above, C denotes various constants independent of ε.

Combining (4.35), (4.36), we establish the following result:

Theorem 4.11. The pair $[y^\varepsilon, u_\varepsilon]$ is suboptimal for the problem (4.26) in the sense that:

i) it satisfies (4.27)-(4.30),

ii) $g(y^\varepsilon) + h(u_\varepsilon) \to g(y^*) + h(u^*)$,

iii) $\mathrm{dist}(y^\varepsilon(t,x), [a,b]) \le C\varepsilon, \quad \varepsilon > 0.$

Remark 4.12. A more detailed discussion along these lines and some numerical examples may be found in [130], [116]. Generally speaking, from a numerical point of view, the variational inequality method is stable with respect to rough initial guesses for the first iteration of the algorithm.

Now, we turn to the question of the equivalence between state constrained control problems and problems governed by variational inequalities (without state constraints). Such equivalence properties appear in certain situations and we give some examples of this type.

First we analyse the admissibility problem. In many applications, when both state and control constraints are given, it is a difficult task to find an admissible pair, or at least an approximate admissible pair, in a sense to be precised.

We define (S) to be a constrained system given by (4.2), (4.3), (4.4) and

(4.37) $\qquad u \in U_0.$

The hypotheses on A, B, f, y_0, C, U_0 are stated at the begining of this section.

Since C and U_0 are closed subsets, they may be very "thin" and it is possible that (S) has no admissible pair. We take:

(4.38) $\qquad C_\delta = \{v \in H;\ \mathrm{dist}(v,C)_H \le \delta\},$

(4.39) $\qquad U_\delta = \{u \in L^2(0,T;U);\ \mathrm{dist}(u,U_0)_{L^2(0,T;U)} \le \delta\}.$

These are convex, closed subsets with nonempty interior. We denote (S_δ) the system given by (4.2), (4.3) and $u \in U_\delta$, $y \in C_\delta$. An admissible pair for (S_δ) is called δ-admissible for the system (S). We relax the admissibility hypothesis from the beginning of this section to

(4.40) (S) has a δ-admissible pair $[y_\delta, u_\delta]$.

Let $\delta > 0$ be fixed and $\Psi : L^2(0,T;U) \to]-\infty,+\infty]$, $\alpha : H \to]-\infty,+\infty]$ be the indicator functions of U_δ, C_δ. Obviously, (S_δ) may be rewritten as a control problem as follows

(4.41) Minimize $\Psi(u)$

for $u \in L^2(0,T;U)$, y given by (4.2), (4.3) and $y(t) \in C_\delta$, $t \in [0,T]$.

We associate with (4.41) the problem governed by variational inequalities (without state constraints)

(4.42) Minimize$\left\{ \Psi(u) + 1/2 \, |w|^2_{L^2(0,T;V^*)} \right\}$,

(4.43) $y' + Ay + w = Bu + f$, $w \in \partial\alpha(y)$

and (4.3).

An elementary argument establishes the equivalence result:

Proposition 4.13. The set of δ- admissible pairs for the system (S) coincides with the set of optimal pairs for the problem (4.42), (4.43).

Remark 4.14. It is not necessary to introduce a new parameter ε in (4.43), as in (4.10), since we prove equivalence, not only approximation. Now, it is possible to apply a smoothing technique in (4.43). In this way, one may find numerically δ' - admissible pairs $(\delta' > \delta)$ for the system (S).

The equivalence property is valid in a general setting. We return to the problem (4.1)-(4.4) and we suppose that $\dot{C}$ is bounded in H. Let $B^* : H \to U$ be the adjoint of B (H and U are identified with their duals). It yields that $B^*(C)$ is a closed, convex subset in U.

Consider $\tilde{C} = \{ v \in H;\ B^* v \in B^*(C) \}$ which also is a convex , closed subset in H, $\tilde{C} \supseteq C$. Denote by φ, ζ, ξ the indicator functions of C, $\tilde{C}$, $B^*(C)$ in H, respectively in U. We define the problems:

(4.44) Minimize$\{ g(y) + h(u) \}$

for y,u satisfying to (4.2), (4.3) and the constraint

(4.45) $y(t) \in \tilde{C}$, $t \in [0,T]$;

respectively

(4.46) Minimize$\left\{ g(y) + h(u-w) + 1/2 \, |w|^2_{L^2(0,T;U)} \right\}$,

(4.47) $\quad y' + Ay + Bw = Bu + f, \quad w \in \partial\xi(B^*y),$

and (4.3).

Remark 4.15. Let $P : H \to H$ be the monotone operator, possibly multivalued, $Py = B\partial\xi(B^*y)$. We have $Py \subset \partial\zeta(y)$ with equality if some interiority assumptions are imposed (see §3.1, Chapter I) because $\zeta = \xi \circ B^*$. By monotonicity, the equation (4.47) has at most one solution. Any solution of (4.47) also satisfies:

(4.48) $\quad y' + Ay + \partial\zeta(y) \ni Bu + f,$

but the converse is valid only if the section of $\partial\zeta(y)$, for which (4.48) is fulfilled, also belongs to Py.

Therefore the problem (4.46), (4.47) is a singular control problem since we haven't existence in (4.47) for any u and, even in case of existence it is possible that $w \notin L^2(0,T;U)$.

The set of admissible controls for (4.46), (4.47) contains all the admissible controls for the problem (4.1)-(4.4) with the same cost because we may take $w = 0$.

It is possible to prove the existence of the optimal pairs for the problems (4.44), (4.46) in a standard manner.

Theorem 4.16. *The problems (4.46) and (4.44) are equivalent in the sense that they have the same optimal pairs and the same optimal value.*

Proof

If $[y^*,u^*]$ is a solution of (4.44), then $B^*y^* \in B^*(C)$ and $0 \in \partial\xi(B^*y^*)$. It yields that $[y^*,u^*]$ is an admissible pair for (4.46) with $w^* = 0$.

We denote $\tilde{J}, J_1$ the cost functionals (4.44), respectively (4.46) and we have

$$\tilde{J}(y^*,u^*) = J_1(y^*,u^*) \geq J_1(\hat{y},\hat{u}),$$

where $[\hat{y},\hat{u}]$ is an optimal pair for the problem (4.46). Obviously, by (4.47) we have $B^*\hat{y} \in B^*(C)$, so $\hat{y} \in \tilde{C}$ and this shows that $[\hat{y},\hat{u}-\hat{w}]$ is admissible for (4.44), where $\hat{w} \in \partial\xi(B^*y)$ is given by (4.47). Moreover

$$\tilde{J}(\hat{y},\hat{u} - \hat{w}) \leq J_1(\hat{y},\hat{u}) \leq J_1(y^*,u^*) = \tilde{J}(y^*,u^*).$$

It yields that $\hat{w} = 0$ and $[\hat{y},\hat{u}]$ is an optimal pair for the problem (4.44).

Corollary 4.17. *If $C = \tilde{C}$ then the problem (4.1) - (4.4) is equivalent with the problem (4.46), (4.47).*

Remark 4.18. If B is onto (distributed control), by the alternative theorem, B^* is injective and, consequently, $C = \tilde{C}$. In many situations this equality is valid in an implicit manner and we give an example of a boundary control problem, according to Bonnans and Tiba [132].

Let Ω be a bounded domain in R^N and $V = H^1(\Omega)$, $H = L^2(\Omega)$, $A : H^1(\Omega) \to H^1(\Omega)^*$ is the Laplace operator. We define $B : L^2(\Omega) \to H^1(\Omega)^*$ by

(4.49) $$(Bu,v)_{V \times V^*} = \int_\Gamma uv d\sigma, \qquad v \in V.$$

We remark, that $B^* : V \to L^2(\Gamma)$ is given by $B^* v = v|_\Gamma$ for $v \in V$. We consider the following control problem, where $y_d \in L^2(Q)$:

(4.50) $$\text{Minimize}\left\{1/2\int_0^T\int_\Omega (y - y_d)^2 dxdt + 1/2\int_0^T\int_\Gamma u^2 d\sigma dt\right. ,$$

(4.51) $\partial y/\partial t - \Delta y = 0$ in $\Omega \times]0,T[$,

(4.52) $y(0,x) = y_0(x)$ in Ω,

(4.53) $\partial y/\partial n = u$ in Σ,

with state constraints

(4.54) $-m \leq y|_\Gamma \leq m$, $t \in [0,T]$.

Therefore, by definition, we have

$$C = \{ v \in H^1(\Omega); -m \leq v|_\Gamma \leq m\} = \{v \in H^1(\Omega); -m \leq B^* v \leq m\} = \tilde{C}.$$

Then, the problem (4.40) is equivalent with the problem (4.46), which, on this example, becomes:

(4.55) $$\text{Minimize}\left\{1/2\int_0^T\int_\Omega (y - y_d)^2 dxdt + 1/2\int_\Sigma w^2 d\sigma dt + \frac{1}{2}\int_\Sigma (u-w)^2 d\sigma dt\right\},$$

(4.56) $\partial y/\partial t - \Delta y = 0$ in $\Omega \times]0,T[$,

(4.57) $y(0,x) = y_0(x)$ in Ω,

(4.58) $\partial y/\partial n = u-w$, $w \in \partial\gamma(y|_\Gamma)$,

where $\gamma : L^2(\Gamma) \to]-\infty,+\infty]$ is given by

(4.59) $$\gamma(y) = \begin{cases} 0 & -m \leq y \leq m \quad \text{a.e.} \Gamma, \\ +\infty & \text{otherwise.} \end{cases}$$

The problem (4.55)-(4.58) is governed by variational inequalities with unilateral conditions on the boundary. If control constraints $u \in U_{ad}$ are imposed in (4.50), they become $u - w \in U_{ad}$ in (4.55).

Remark 4.19. In this example, we have $B : U \to V^*$ linear, continuous. This is a direct extension of the assumptions of Thm. 4.16, which includes the case of boundary control.

We close this section with another equivalence result, which plays a key role in the proof of the optimality conditions for state constrained control problems governed by variational inequalities. As an application we derive the existence of a bang-bang control.

We consider the problem

(4.60) Minimize $\{g(y) + h(u)\}$,

(4.61) $\partial y/\partial t - \Delta y + \beta(y) \ni u$ in $\Omega\times]0,T[$,

(4.62) $y(0,x) = y_0(x)$ in Ω,

(4.63) $y(t,x) = 0$ in Σ

with the constraints

(4.64) $y(t,x) \leq b$ a.e. $\Omega\times]0,T[$,

(4.65) $c \leq u(t,x) \leq d$ a.e. $\Omega\times]0,T[$.

Here g, h, y_0 are as in the problem (4.1) and $\beta \subset R \times R$ is the maximal monotone graph

(4.66)
$$\beta(y) = \begin{cases}]-\infty,0] & y = a, \\ 0 & y > a, \\ \emptyset & y < a. \end{cases}$$

We suppose that a,b,c,d are real constants such that $0\in[c,d]$ (the zero control is accepted) and $0\in[a,b]$ (compatibility with the Dirichlet conditions). We also need that $y_0(x)\in[a,b]$ a.e.Ω and $y_0\in W_0^{1,p}(\Omega)$ with $p > (N+2)/2$. By Thm.4.5, Chapter I, it yields that the unique solution y of the parabolic variational inequality (4.61)-(4.63) satisfies $y\in W^{2,1,p}(Q)$.

Let $\alpha \subset R \times R$ be the maximal monotone graph given by

(4.67)
$$\alpha(y) = \begin{cases} 0 & y < b, \\ [0,+\infty[& y = b, \\ \emptyset & y > b. \end{cases}$$

Obviously $\alpha+\beta\subset R \times R$ is maximal monotone.

We associate with (4.60) - (4.65) the problem

(4.68) Minimize $\{g(y) + h(u)\}$

$\partial y/\partial t - \Delta y + \beta(y) + \alpha(y) \ni u$ in $\Omega\times]0,T[$,

$y(0,x) = y_0(x)$ in Ω,

$y(t,x) = 0$ in Σ.

The following lemma plays a fundamental role and also has its own interest.

Lemma 4.20. We assume that $h(u) = f(|u|)$ ($|\cdot|$ is the modulus) and $f : L^2(Q)\to R$ is increasing with respect to the a.e. order, on the positive cone. Then, there exists an optimal pair for the problem (4.68), $[y^*,u^*]$, such that $\alpha(y^*) = 0$ a.e. Q.

Proof

If $y^*(t,x) < b$, then $\alpha(y^*(t,x)) = 0$ by (4.67). If $y^*(t,x) = 0$, then $\beta(y^*(t,x)) = 0$ by (4.66) and, due to the regularity of y^*, it yields $\partial y^*/\partial t - \Delta y^* = 0$. Therefore, on the subset where $y^*(t,x) = b$, we have $u^* = \alpha(y^*) \geq 0$. But, replacing both $\alpha(y^*)$ and u^* by 0 on this subset, the cost would decrease (by our hypothesis) since y^* remains unchanged.

Remark 4.21. If f is strictly increasing, in the sense that $0 \leq u \leq v$ a.e.Q and $u \neq v$ implies $f(u) < f(v)$, then all the optimal pairs have this extremality property.

Remark 4.22. Similarly, in the problem (4.60) one may establish the existence of an optimal pair $[\hat{y},\hat{u}]$ with $\beta(\hat{y}) = 0$ a.e. Q.

Theorem 4.23. If f is strictly increasing, then the problems (4.60) and (4.68) are equivalent. If f is increasing, then any solution of (4.60) is solution of (4.68) too.

Proof.

Let $[y^*,u^*]$ be an optimal pair for (4.68). Obviously $y^*(t,x) \leq b$ a.e. $\Omega \times]0,T[$ and Lemma 4.20 shows that $[y^*,u^*]$ is admissible for (4.60). Let J and J_a be the cost functionals of (4.60) and (4.68). Then

$$J(y^*,u^*) = J_a(y^*,u^*) \geq J(\hat{y},\hat{u}),$$

where $[\hat{y},\hat{u}]$ is an optimal pair for (4.60).

Since $\hat{y} \leq b$ a.e.Q, we may take $\alpha(\hat{y}) = 0$, so $[\hat{y},\hat{u}]$ is admissible for (4.68) with the same cost

$$J_a(\hat{y},\hat{u}) = J(\hat{y},\hat{u}) \geq J_a(y^*,u^*).$$

It yields that (4.60) and (4.68) have the same optimal value and optimal pairs.

Remark 4.24. Under the above hypotheses on h we see that the problem (4.60) is in fact equivalent with

$$\text{(4.69)} \qquad \begin{aligned} &\text{Minimize}\left\{g(y) + h(u)\right\}, && \\ &\partial y/\partial t - \Delta y = u && \text{in } Q,\\ &y(0,x) = y_0(x) && \text{in } \Omega,\\ &y(t,x) = 0 && \text{in } \Sigma,\\ &a \leq y(t,x) \leq b && \text{in } Q,\\ &c \leq u(t,x) \leq d && \text{in } Q. \end{aligned}$$

This proves the rather surprising conclusion that, in fact, the optimization problem (4.60) governed by the variational inequality (4.61) is equivalent with a convex minimization problem.

Corollary 4.25. If g or h are strictly convex, then the problem (4.60)-(4.65) has a unique optimal pair.

This may be inferred by Remark 4.22 and the strict convexity of the problem (4.58).

We return to the problem (4.68). In the same way as in §1, we obtain the optimality conditions for (4.68), therefore for (4.60) too. For the sake of simplicity, we take h = 0.

Theorem 4.26. There are $p^* \in L^\infty(0,T;L^2(\Omega)) \cap L^2(0,T;H_0^1(\Omega))$ and $\rho \in \mathcal{D}'(Q)$, $\operatorname{supp}\rho \subset M = \{(t,x) \in Q;\ y^*(t,x) \in \{a,b\}\}$ (the coincidence set) such that

(4.70) $\quad \partial p^*/\partial t + \Delta p^* - \rho = \partial g(y^*),$

(4.71) $\quad \int_Q p^*(u^* - v) \geq 0,\ \forall v \in L^2(Q),\ c \leq v \leq d$ a.e. Q.

Moreover $p^* \in W^{2,1,p}_{loc}(E)$, where $E = Q - M$ is an open subset in Q.

Here, we use the condition $p \geq (N+2)/2$ essentially.

Now, we are able to obtain the structure of an optimal control for the problem (4.60). By Lemma 4.20 we see that $u^* = 0$ a.e.M. An appropriate choice of v in (4.71) gives

$$\int_E p^*(u^* - v)dxdt \geq 0,\ \forall v \in L^2(E),\ c \leq v(t,x) \leq d.$$

Obviously $u^* = d$ a.e. in $\{(t,x) \in E;\ p^*(t,x) > 0\}$ and $u^* = c$ a.e. in $\{(t,x) \in E;\ p^*(t,x) < 0\}$.

Let us consider $g(y) = 1/2|y|^2_{L^2(0,T;H)}$. If the set $\{(t,x) \in E;\ p^*(t,x) = 0\}$ has a positive measure, then (4.70) implies (together with the regularity of p^*) that $y^* = 0$ on this set. Using again Lemma 4.20 and (4.68) it yields that $u^* = 0$ on this subset. Then we have:

Corollary 4.27. If h = 0 and $g(y) = 1/2|y|^2_{L^2(0,T;H)}$, then the problem (4.60)-(4.65) has an optimal control u^* which is bang-bang.

We shortly comment another example, of a different nature. We take

$$h(u) = \int_Q u\,dxdt$$

and c = 0, d > 0 in the definition of U_{ad}. Then we may take f(u) = h(u) since h(u) = f(|u|) on U_{ad}. This f is strictly increasing on the positive cone in $L^2(Q)$ and the problems (4.60) and (4.68) are equivalent. We fix

$$g(y) = \int_Q y\,dxdt$$

in order to avoid strict convexity and the uniqueness of the optimal pair. By (4.70) we see that the set $\{(t,x) \in E;\ p^*(t,x) = 0\}$ has zero measure. Therefore, on E, every optimal control is bang-bang. Moreover, obviously, on M we have $u^* = 0$. We obtain that all the optimal controls take only the values 0 and d a.e.Q.

Remark 4.28. It is possible to extend these results to the more general situation discussed in §1, when constraints are imposed.

5. Elliptic and optimal design problems

5.1. Optimal shape design

The aim of this section is to discuss an optimal design problem, by the variational inequality method in order to argue the usefulness of this approach. We fix our attention on the design of the optimal covering of an obstacle, the so called "packaging problem", Zolesio, Sokolowski, Benedict [142], Haslinger and Neittaanmaki [59], Ch. X.

This is quite a challenging control problem due to several difficulties: it is governed by variational inequalities, it is an optimal design problem, it involves special state constraints.

Consider a membrane $\Omega(\alpha)$ in possible contact with a rigid obstacle G. Let φ describe the shape of the obstacle and $\Omega(\alpha)$ be given by

$$(5.1)\qquad \Omega(\alpha)=\{(x_1,x_2)\in R^2;\ x_2\in]0,1[,\ 0<x_1<\alpha(x_2)\},$$

where $\alpha\in U_{ad}$ is the function describing the moving part $\Gamma(\alpha)$ of the boundary $\partial\Omega(\alpha)$:

$$\Gamma(\alpha)=\{(x_1,x_2);\ x_1=\alpha(x_2),\ x_2\in]0,1[\},$$
$$U_{ad}=\{\alpha\in W^{1,\infty}(0,1);\ a\le\alpha\le b,|\alpha'|\le c\},$$

with a,b,c positive constants such that $U_{ad}\neq\emptyset$.

On any $\Omega(\alpha)$, $\alpha\in U_{ad}$, we introduce the following variational inequality: find $u(\alpha)\in K(\Omega(\alpha))$ such that

$$(5.2)\qquad \begin{aligned}&(\text{grad } u(\alpha), \text{grad}(v-u(\alpha)))_{L^2(\Omega(\alpha))}\ge(f,v-u(\alpha))_{L^2(\Omega(\alpha))}\\ &\forall v\in K(\Omega(\alpha))=\{v\in H^1_0(\Omega(\alpha));v\ge\varphi \text{ a.e.} \Omega(\alpha)\}.\end{aligned}$$

We assume that $f\in L^2(\hat\Omega)$, $\hat\Omega=]0,b[\times]0,1[$, and $\varphi\in H^1(\hat\Omega)$, $\varphi\le 0$ on $\partial\hat\Omega$ and in $[a,b]\times[0,1]$.

The equation (5.2) describes the vertical displacement $u(\alpha)$ of the membrane $\Omega(\alpha)$ (the equilibrium position) under the load f and in contact with the obstacle G. Formally, (5.2) may be rewritten as:

$$\begin{aligned}&-\Delta u(\alpha)\ge f && \text{in}\,\Omega(\alpha),\\ &u(\alpha)\ge\varphi && \text{in}\,\Omega(\alpha),\\ &(-\Delta u(\alpha)-f)(u-\varphi)=0 && \text{in}\,\Omega(\alpha),\\ &u(\alpha)=0 && \text{in}\,\partial\Omega(\alpha).\end{aligned}$$

We denote

$$Z(u(\alpha))=\{x\in\Omega(\alpha);u(\alpha)(x)=\varphi(x)\},$$

the contact or the coincidence region, and the packaging problem consists in minimizing the area of $\Omega(\alpha)$ such that the contact region $Z(u(\alpha))$ of the solution of (5.2) contains a given subset $\Omega_0 \subset \hat{\Omega}$. That is, we consider the optimization problem

(5.3) $\quad$ Minimize $\int_0^1 \alpha(x)dx$

for $\alpha \in U_{ad}$ and such that $u(\alpha)$, the solution of (5.2) corresponding to α, satisfies the constraint

(5.4) $\quad$ $\Omega_0 \subset Z(u(\alpha))$.

Therefore, the minimization parameter in the problem (5.3) is $\alpha \in U_{ad}$, that is, the minimization is taken with respect to the family of domains $\Omega(\alpha)$ given by (5.1). This is specific to optimal design problems, where, roughly speaking, one wants to find the optimal shape of a structure with respect to a certain cost. The following continuity result is taken from Haslinger and Neittaanmaki [59] and plays an important role in the sequel. If $\alpha \in U_{ad}$ we write $\tilde{u}$ for the extension of $u(\alpha)$ to $\hat{\Omega}$ by zero.

Theorem 5.1. Let $\alpha_n \to \alpha$ uniformly in $[0,1]$ and let $u_n = u(\alpha_n)$ be the solutions of (5.2). Then, there exists a subsequence, denoted again u_n, such that $\tilde{u}_n \to U$ strongly in $H^1(\hat{\Omega})$ and $U|_{\Omega(\alpha)}$ is the solution of (5.2) corresponding to α.

Proof

One may easily infer that $|\tilde{u}_n|_{H_0^1(\hat{\Omega})} \leq C$ and, by taking subsequences, that $\tilde{u}_n \to U$ weakly in $H_0^1(\hat{\Omega})$. Since $\alpha_n \to \alpha$ uniformly in $[0,1]$ we see that $U|_{\hat{\Omega} \setminus \Omega(\alpha)} = 0$, so $U|_{\Omega(\alpha)} \in H_0^1(\Omega(\alpha))$. Moreover $\tilde{u}_n \geq \varphi$ a.e. $\hat{\Omega}$ and we obtain $U \geq \varphi$ a.e. $\hat{\Omega}$, that is $U \in K(\Omega(\alpha))$.

For any $v \in K(\Omega(\alpha))$ there exist a subsequence $\{\alpha_n\}$ and a sequence $v_n \in H_0^1(\hat{\Omega})$ such that

(5.5) $\quad$ $v_n \to \tilde{v}$ $\quad$ in $H_0^1(\hat{\Omega})$,

(5.6) $\quad$ $v_n|_{\Omega(\alpha_n)} \in K(\Omega(\alpha_n))$.

The sequence $\{v_n\}$ may be constructed as follows. There is $\{\omega_k\} \subset \mathcal{D}(\Omega(\alpha))$ such that $\omega_k \to v$ in $H_0^1(\Omega(\alpha))$ and $\tilde{\omega}_k \to \tilde{v}$ in $H_0^1(\hat{\Omega})$. Let $v_k = \sup(\tilde{\omega}_k, \varphi)$. Obviously $v_k \in H_0^1(\hat{\Omega})$ and $v_k \geq \varphi$ a.e. in $\hat{\Omega}$. By the continuity of the sup $(.,.)$ application with respect to the $H^1(\hat{\Omega})$ norm, we get (5.5).

Let k_0 be fixed and $G_{k_0} = \operatorname{supp} \omega_{k_0}$. As $\alpha_n \to \alpha$ uniformly, there is $n_0 = n(k_0)$ such that $\Omega(\alpha_{n_0}) \supset G_{k_0}$ and $v_{k_0} = 0$ on $\hat{\Omega} - \Omega(\alpha_{n_0})$. Hence $v_{k_0}|_{\Omega(\alpha_{n_0})} \in K(\Omega(\alpha_{n_0}))$ and (5.6) follows too.

From the definition of $u(\alpha_n)$ as the solution of (5.2) and (5.6) we obtain

$$(\operatorname{grad} u_n, \operatorname{grad}(v_n - u_n))_{L^2(\Omega(\alpha_n))} \geq (f, v_n - u_n)_{L^2(\Omega(\alpha_n))},$$

$$(\operatorname{grad} u_n, \operatorname{grad}(v_n - u_n))_{L^2(\hat{\Omega})} \geq (f, v_n - \tilde{u}_n)_{L^2(\hat{\Omega})}.$$

One may pass to the limit and deduce:

$$(\text{grad } U, \text{grad}(\tilde{v}-U))_{L^2(\hat{\Omega})} \geq (f, \tilde{v}-U)_{L^2(\hat{\Omega})},$$
$$(\text{grad } u, \text{grad}(v-u))_{L^2(\Omega(\alpha))} \geq (f, v-u)_{L^2(\Omega(\alpha))},$$

where $u = U|_{\Omega(\alpha)}$. Because $v \in K(\Omega(\alpha))$ is arbitrary, we see that u solves (5.2), so $u = u(\alpha)$.

Finally, we show that, on a subsequence, $\tilde{u}_n \to U$ in the norm of $H^1(\hat{\Omega})$. We apply (5.5), (5.6) to U and we denote w_n the obtained sequence. We have

$$0 \leq d|U - \tilde{u}_n|^2_{H^1(\hat{\Omega})} \leq (\text{grad}(U - \tilde{u}_n), \text{grad}(U - \tilde{u}_n)_{L^2(\hat{\Omega})} = (\text{grad } U, \text{grad}(U - \tilde{u}_n))_{L^2(\hat{\Omega})} -$$

$$- (\text{grad } u_n, \text{grad}(U - w_n))_{L^2(\hat{\Omega})} - (\text{grad } u_n, \text{grad}(w_n - u_n))_{L^2(\hat{\Omega})} \leq (\text{grad } U, \text{grad}(U -$$

$$- u_n))_{L^2(\hat{\Omega})} - (\text{grad } u_n, \text{grad}(U - w_n))_{L^2(\hat{\Omega})} - (f, w_n - u_n)_{L^2(\hat{\Omega})} \to 0.$$

Remark 5.2. Since the solution of (5.2) is unique, we obtain the convergence of the whole sequence u_n in Thm.5.1.

Remark 5.3. As an easy consequence of Thm.5.1 and of the compactness of U_{ad}, one may infer the existence of at least are optimal control α^* for the problem (5.2) - (5.4) under the usual assumption on the existence of at least one admissible pair $[u(\hat{\alpha}), \hat{\alpha}]$.

One main difficulty in the treatment of the problem (5.2)-(5.4) consists in the variable character of the domain $\Omega(\alpha)$ on which the problem is given. By scaling the domain $\Omega(\alpha)$ such that it becomes fixed, the optimization parameter α appears as a coefficient in the state system. Optimal design problems of this type were discussed in the work of Haslinger, Neittaanmaki and Tiba [131] by the penalization and the variational inequality methods.

In the sequel, we show that it is possible to use the variational inequality method directly in the problem (5.2) - (5.4) and to obtain precise approximation results.

We start with a relaxation of the state constraints. Let $\{\psi_n^\delta\}$ be a family of smooth functions on $\hat{\Omega}$ satisfying the conditions

(5.7) $\quad \psi_n^\delta \leq 1/n + \varphi \quad$ in $\Omega_{0,n}$,

(5.8) $\quad \psi_n^\delta \geq n \quad$ in $\Omega(\alpha) - \Omega_0^n$,

(5.9) $\quad \psi_n^\delta \leq \delta + \varphi \quad$ in Ω_0, $\psi_n^\delta = \delta + \varphi \quad$ in $\partial\Omega_0$,

(5.10) $\quad \psi_n^\delta \geq \psi_{n-1}^\delta \quad$ in $\Omega - \Omega_0$,

(5.11) for any continuous function u on $\Omega(\alpha)$, with $u|_{\Omega_0} = \varphi$, there exists $n_0 \in N$, such that $\psi_n^\delta \geq u$ in $\Omega(\alpha)$ for $n \geq n_0$.

Above, we denote

(5.12) $$\Omega_{0,n} = \{x \in \Omega_0;\ \text{dist}(x, \partial\Omega_0) \geq 1/n\},$$

(5.13) $$\Omega_0^n = \{x \in \Omega(\alpha);\ \text{dist}(x, \Omega_\delta) \leq 1/n\}$$

and we assume that $n \geq 1/\delta$.

Roughly speaking, the family $\{\psi_n^\delta\}$ is an approximation of the indicator function of Ω_0 plus φ and the conditions (5.7) - (5.11) may be viewed as regularity assumptions on Ω_0. See the end of this section for an effective construction of such a family.

For the sake of simplicity, we put $\varphi = 0$.

Lemma 5.4. For any admissible pair $[u(\hat{\alpha}),\hat{\alpha}]$ of the problem (5.2) - (5.4), there exists $\bar{n}$ such that $[u(\hat{\alpha}),\hat{\alpha}]$ is an admissible pair for the approximating problem (5.2), (5.3) and

$$(5.14) \qquad u \leq \psi_n^\delta \qquad \text{a.e. } \Omega(\alpha).$$

Proof

For $f \in L^2(\hat{\Omega})$, since we work in space dimension 2, it is wellknown that the solution of the variational inequality (5.2) is continuous in $\Omega(\hat{\alpha})$. As $[u(\hat{\alpha}),\hat{\alpha}]$ is an admissible pair, we have $u = 0$ on Ω_0 and, by (5.11), there is $\bar{n} \in N$ such that $u(\hat{\alpha}) \leq \psi_n^\delta$ for $n \geq \bar{n}$ and (5.14) is fulfilled.

Remark 5.5. In particular, Lemma 5.4 is valid for any optimal pair $[u(\alpha^*),\alpha^*]$.

Remark 5.6. Using again Thm.5.1 and the compactness of U_{ad}, one may easily establish the existence of at least one optimal pair for the approximating optimal design problem, which we denote $[u(\alpha_n^\delta),\alpha_n^\delta]$.

Proposition 5.7. For $n \to \infty$, on a subsequence, we have $\alpha_n^\delta \to \alpha^\delta$ uniformly and α^δ is an optimal control for the problem (5.2) - (5.4).

Proof

Let J denote the cost functional (5.3). By Lemma 5.4 α^* is admissible for the approximating problem for $n \geq \bar{n}$ and we get

$$(5.15) \qquad J(\alpha_n^\delta) \leq J(\alpha^*).$$

Moreover, as U_{ad} is compact in $C(0,1)$, we may assume that, on a subsequence, $\alpha_n^\delta \to \alpha^\delta \in U_{ad}$, uniformly in $[0,1]$ and (5.15) gives

$$(5.16) \qquad J(\alpha^\delta) \leq J(\alpha^*).$$

Then, Thm.5.1 implies that $\tilde{u}(\alpha_n^\delta) \to \tilde{u}(\alpha^\delta)$ strongly in $H_0^1(\hat{\Omega})$. By (5.7) we see that $\tilde{u}(\alpha^\delta) = 0$ a.e. Ω_0, that is the pair $[u(\alpha^\delta), \alpha^\delta]$ is admissible for the problem (5.2) - (5.4). From (5.16) we see that it is optimal.

Remark 5.8. For n sufficiently large, the pairs $[u(\alpha_n^\delta), \alpha_n^\delta]$ satisfy:

- $J(\alpha_n^\delta) \leq J(\alpha^*)$,
- the state equation,
- $\alpha_n^\delta \in U_{ad}$,
- $0 \leq u(\alpha_n^\delta) \leq \delta$ in Ω_0 (pointwise estimate).

Therefore they are suboptimal for the problem (5.2) - (5.4).

In order to remove the state constraint (5.14) we apply the variational inequality technique. We consider the problem without state constraints:

(5.17) $\quad \text{Minimize}\{J(\alpha) + 1/\varepsilon\, |\gamma_\varepsilon(u-\psi_n^\delta)|^2_{L^2(\Omega(\alpha))}\}$,

(5.18) $\quad -\Delta u + \beta(u) + \gamma_\varepsilon(u-\psi_n^\delta) \ni f$,

(5.19) $\quad u|_{\partial\Omega(\alpha)} = 0$,

(5.20) $\quad \alpha \in U_{ad}$.

Here γ_ε is the Yosida approximation of the maximal monotone graph

$$\gamma(y) = \begin{cases} 0 & y < 0, \\ ]0,\infty[& y = 0, \\ \phi & y > 0, \end{cases}$$

and β is given by (3.3).

Lemma 5.9. Assume that $f \in L^\infty(\hat{\Omega})$, then

(5.21) $\quad |\gamma_\varepsilon(u-\psi_n^\delta)|_{C(\overline{\Omega(\alpha)})} \le |f + \Delta\psi_n^\delta|_{L^\infty(\Omega(\alpha))}$,

where u is the solution of (5.18), (5.19).

Proof

By the regularity of u we know that $\gamma_\varepsilon(u-\psi_n^\delta) \in C(\overline{\Omega(\alpha)})$ since γ_ε is Lipschitzian. Moreover, as $\gamma_\varepsilon(y) = 0$ for $y \le 0$, we get

(5.22) $\quad \beta(u)\gamma_\varepsilon(u-\psi_n^\delta) = 0 \qquad \text{in } \Omega(\alpha).$

Multiply (5.18) by $\gamma_\varepsilon^{p-1}(u-\psi_n^\delta)$, $p > 2$ even. We have:

$$-\int_{\Omega(\alpha)} \Delta u \gamma_\varepsilon^{p-1}(u-\psi_n^\delta)dx = -\int_{\Omega(\alpha)} \Delta(u-\psi_n^\delta)\gamma_\varepsilon^{p-1}(u-\psi_n^\delta)dx - \int_{\Omega(\alpha)} \Delta\psi_n^\delta \gamma_\varepsilon^{p-1}(u-\psi_n^\delta)dx =$$

$$= \int_{\Omega(\alpha)} \text{grad}(u-\psi_n^\delta)\text{grad}\gamma_\varepsilon^{p-1}(u-\psi_n^\delta)dx - \int_{\partial\Omega(\alpha)} \partial/\partial n(u-\psi_n^\delta)\gamma_\varepsilon^{p-1}(-\psi_n^\delta)d\sigma -$$

$$-\int_{\Omega(\alpha)} \Delta\psi_n^\delta \gamma_\varepsilon^{p-1}(u-\psi_n^\delta)dx \ge -\int_{\Omega(\alpha)} \Delta\psi_n^\delta \gamma_\varepsilon^{p-1}(u-\psi_n^\delta)dx.$$

Combining this with (5.22) we infer the estimate

$$\int_{\Omega(\alpha)} |\gamma_\varepsilon(u-\psi_n^\delta)|^p dx \le \int_{\Omega(\alpha)} |f + \Delta\psi_n^\delta|\, |\gamma_\varepsilon(u-\psi_n^\delta)|^{p-1} dx$$

for all $p > 2$ even. Then

$$|\gamma_\varepsilon(u-\psi_n^\delta)|_{L^p(\Omega(\alpha))} \leq |f+\Delta\psi_n^\delta|_{L^p(\Omega(\alpha))}$$

and passing to the limit $p\to\infty$, we prove (5.21).

Corollary 5.10. We have

$$\sup_{x\in\bar{\Omega}_0} p(u-\psi_n^\delta)_+ \leq \varepsilon|f+\Delta\psi_n^\delta|_{L^\infty(\Omega(\alpha))}. \tag{5.23}$$

Proof

This follows by (5.21) and the properties of γ_ε.

Remark 5.11. Obviously $u \geq 0$ as the solution of (5.18). Therefore, on Ω_0, we have the pointwise estimate

$$0 \leq u \leq \psi_n^\delta + |f+\Delta\psi_n^\delta|.\varepsilon \leq \delta + \varepsilon|f+\Delta\psi_n^\delta|_{L^\infty(\hat{\Omega})}.$$

Remark 5.12. An argument similar to Thm.5.1 shows that the problem (5.17)-(5.20) has at least one optimal pair, which we denote $[u(\alpha_{n,\varepsilon}^\delta), \alpha_{n,\varepsilon}^\delta]$. Moreover, since $\gamma_\varepsilon(y) = 0$ for $y \leq 0$, then any admissible pair for (5.2), (5.3), (5.14) is also admissible for (5.17) - (5.20) with the same cost. It yields

$$J(\alpha_{n,\varepsilon}^\delta) + 1/\varepsilon\,|\gamma_\varepsilon(u(\alpha_{n,\varepsilon}^\delta)-\psi_n^\delta)|^2_{L^2(\Omega(\alpha_{n,\varepsilon}^\delta))} \leq J(\alpha^*), \tag{5.24}$$

for n sufficiently large.

By (5.23), (5.24) we see that the pair $[u(\alpha_{n,\varepsilon}^\delta), \alpha_{n,\varepsilon}^\delta]$ has nice properties. However, it doesn't satisfy the state equation (5.2) and this may cause troubles. In order to check this, we denote shortly by u^ε the solution of (5.2) corresponding to $\alpha_{n,\varepsilon}^\delta$.

Assume now that the cost functional (5.17) contains the term

$$1/\varepsilon\;|\gamma_\varepsilon(u-\psi_n^\delta)|^2_{L^q(\Omega(\alpha))}$$

with some $q > 2$. All the above results remain true and, by (5.24), we have

$$|\gamma_\varepsilon(u(\alpha_{n,\varepsilon}^\delta)-\psi_n^\delta)|_{L^q(\Omega(\alpha_{n,\varepsilon}^\delta))} \leq C\varepsilon^{1/2}$$

with C independent of n,δ,ε.

It is possible to apply a result on the Lipschitzian dependence of the solution of variational inequalities with respect to the right-hand side, due to Brezis [20], p.28. As we work in space dimension 2, we obtain

$$|u(\alpha_{n,\varepsilon}^\delta)-u^\varepsilon|_{L^\infty(\Omega(\alpha_{n,\varepsilon}^\delta))} \leq C\,|\gamma_\varepsilon(u(\alpha_{n,\varepsilon}^\delta)-\psi_n^\delta)|_{L^q(\Omega(\alpha_{n,\varepsilon}^\delta))}.$$

The following result is proved

Corollary 5.13. The pair $[\alpha^{\delta}_{n,\varepsilon}, u^{\varepsilon}]$ is suboptimal for the problem (5.2) - (5.4) in the following sense:

i) it satisfies the state equation,

ii) $J(\alpha^{\delta}_{n,\varepsilon}) \leq J(\alpha^{*})$,

iii) $\alpha^{\delta}_{n,\varepsilon} \in U_{ad}$,

iv) $0 \leq u^{\varepsilon}|_{\Omega_0} \leq \delta + \varepsilon |f + \Delta\psi^{\delta}_n|_{L^{\infty}(\Omega(\alpha^{\delta}_{n,\varepsilon}))} + C\varepsilon^{1/2}$.

We close this section with an example of a family $\{\psi^{\delta}_n\}$ satisfying (5.7) - (5.11). We discuss the case $\Omega(\alpha) \subset R$ and next we study a situation with $\Omega(\alpha) \subset R^2$.

So, let $\Omega(\alpha)$ be an interval of R and Ω_0 be some subinterval of positive length. The situation when Ω_0 is the union of some disjoint subintervals may be studied similarly.

For $n \geq 1/\delta$ we define the sequence $\{\psi^{\delta}_n\}$ such that it satisfies the conditions (5.7) - (5.10) and

$$(5.11)' \qquad |\dot{\psi}^{\delta}_n| \geq n \quad \text{on } \Omega^n_0 - \Omega_0.$$

We have to show that $\{\psi^{\delta}_n\}$ satisfies (5.11) too.

Let u be any continuous function on $\overline{\Omega(\alpha)}$, $u|_{\partial\Omega(\alpha)} = 0$, $u|_{\Omega_0} = 0$ (we have fixed as before $\varphi = 0$). Let u_λ be defined by

$$u_\lambda(x) = \begin{cases} \int_R u(x+\lambda-\lambda\tau)\rho(\tau)d\tau & x \leq x_0, \\ \int_R u(x-\lambda-\lambda\tau)\rho(\tau)d\tau & x \geq x_0, \end{cases}$$

where x_0 is the middle of Ω_0 and ρ is a Friedrich mollifier, that is $\rho \in C^{\infty}_0(R)$, $\rho(x) = 0$ for $|x| \geq 1$, $\rho(-s) = \rho(s)$, $\rho(s) \geq 0$, $\int_{-\infty}^{\infty} \rho(s)ds = 1$. Since $u|_{\Omega_0} = 0$, we get $u_\lambda \in C^{\infty}(\overline{\Omega(\alpha)})$, $u_\lambda|_{\Omega_0} = 0$ and, for sufficiently small λ, we have

$$(5.25) \qquad u_\lambda(x) \geq u(x) - \delta,$$

where $\delta > 0$ is fixed.

As u is bounded on $\overline{\Omega(\alpha)}$, $\{u_\lambda(x)\}$ is uniformly bounded on $\overline{\Omega(\alpha)}$. There is $n_0 \in N$ such that

$$\psi^{\delta}_n \geq u_\lambda + \delta \quad \text{in } \Omega(\alpha) - \Omega^n_0, \qquad n \geq n_0,$$
$$\psi^{\delta}_n \geq u \quad \text{in } \Omega(\alpha) - \Omega^n_0, \qquad n \geq n_0,$$

(n_0 depends only on u).

From $u_\lambda \in C^{\infty}(\overline{\Omega(\alpha)})$, there is $m = \sup|u'_\lambda|$ on $\overline{\Omega(\alpha)}$. We can find n_λ (which depends on λ) such that

(5.26) $|\dot{\psi}_n^\delta| \geq |u'_\lambda|$ on $\Omega_0^n - \Omega_0$.

By (5.9) we have $\psi_n^\delta|_{\partial\Omega_0} = \delta \geq \delta + u_\lambda|_{\partial\Omega_0}$.

This, combined with (5.26), gives

$$\psi_n^\delta \geq \delta + u_\lambda \text{ on } \Omega_0^n - \Omega_0.$$

Of course, (5.26) should be understood in the correct way without the modulus. Finally, (5.25) shows $\psi_n^\delta \geq u$ on $\Omega_0^n - \Omega_0$ and we conclude that $\psi_n^\delta \geq u$ in $\Omega(\alpha)$ for $n \geq n_\lambda$. The above construction may be extended directly to $\Omega(\alpha) \subset R^2$, when Ω_0 is a disc in $\Omega(\alpha)$.

Remark 5.14. Other works dealing with optimal design problems governed by variational inequalities are due to E.Casas [34], Gh.Morosanu and Zheng - Xu He [57], Hlavacek, Bock, Lovisek [58], Zolesio, Payre, Delfour [41], J.Haslinger, P.Neittaanmaki and D.Tiba [131], F.Mignot [74]. We also quote the recent monographs by O.Pirroneau [88], J. Haslinger and P. Neittaanmaki [59] and the volume edited by E.J.Haug and J.Cea [35], devoted to this type of problems.

5.2. Semilinear elliptic variational inequalities

The result of this paragraph may be mainly compared with those from §§ 1-3, this chapter. We examine the optimality conditions for a distributed control problem governed by a semilinear elliptic variational inequality. Comparing with the previous subsection, the minimization parameter is no more the shape of the domain where the partial differential equation is defined.

The obtained adjoint equation is similar in structure, with the parabolic or the hyperbolic case, but the form of the maximum principle is stronger and it is closer to the classical form of the Pontryagin's principle [96]. Moreover, the method used here differs essentially from the other approaches appearing in the book. Mainly, we exhibit a new type of regularization of maximal monotone graphs in R x R which allows uniform approximation of the solutions of variational inequalities. Another important new ingredient in the proof is the Ekeland's variational principle, Ch. I, § 3.3.

By including this subject, we practically cover the entire research area announced in the title and we enlarge the set of methods put at the disposal of the reader.

Let us now state the problem:

(5.27) $$\text{Minimize} \left\{ J(y,u) = \int_\Omega L(x,y(x),u(x))dx \right\}$$

subject to

(5.28) $$Ay + \varphi(x,y(x),u(x)) + \beta(y(x)) \ni 0 \qquad \text{in } \Omega,$$

(5.29) $$y = 0 \qquad \text{in } \partial\Omega,$$

(5.30) $u(x) \in K$ in Ω.

Above $\Omega \subset R^N$ is a bounded domain with smooth boundary $\partial\Omega$, A is a linear differential operator of the form

(5.31) $$Ay = -\sum_{i,j=1}^{N} \partial_{x_i}(a_{ij}(x)\,\partial_{x_j}(y(x)))$$

with $a_{ij} \in C^1(\bar{\Omega})$ and

(5.32) $$\sum_{i,j=1}^{N} a_{ij}(x)\,\xi_i\,\xi_j \geq \omega \sum_{i=1}^{N} \xi_i^2, \qquad \forall\, x \in \Omega,\ \omega > 0,$$

β is a maximal monotone graph in $R \times R$ and $K \subset R$. For the sake of simplicity, we assume that $0 \in \beta(0)$.

We denote by $a : H_0^1(\Omega) \times H_0^1(\Omega) \to R$ the positive bilinear form

$$a(y,z) = \sum_{i,j=1}^{N} \int_{\Omega} a_{ij}(x) \frac{\partial z}{\partial x_j}(x) \frac{\partial y}{\partial x_i}(x)dx\,,$$

associated to the differential operator A.

The mappings $\varphi : \Omega \times R \times K \to R$, $L : \Omega \times R \times K \to R$ are continuous and with continuous derivative with respect to y, such that:

(5.33) $|\varphi(x,0,u)| \leq M(x) + C|u|$,

(5.34) $0 \leq \varphi'_y(x,y,u) \leq [M(x) + C\ u\ ]\ (\ y\ )$,

(5.35) $|L(x,0,u)| \leq M(x) + C|u|$

(5.36) $|L'_y(x,y,u)| \leq [M(x) + C|u|]\eta(|y|)$,

where $M \in L^s(\Omega)$, $s \geq \max[2,N/2]$, C is a positive constant, $\eta : R_+ \to R_+$ is a nondecreasing mapping.

Now, we present the smoothing process of β. We say that, for $\varepsilon > 0$, β_ε is an ε - uniform approximation of β (not to be confused with the Yosida approximation) if it satisfies the following two conditions:

(5.37) $\beta(y+\varepsilon) \geq \beta_\varepsilon(y) \geq \beta(y-\varepsilon)$, $\forall\, y \in R$,

(5.38) $\mathrm{dom}(\beta_\varepsilon) \supset \mathrm{dom}(\beta)$.

Here, we view β and β_ε as multivalued operators extended on R with value $-\infty$ on the left of their domain and $+\infty$ on the right of their domain, and the inequality for sets means

$$\xi \geq \eta \geq \nu\,, \quad \forall\, \xi \in \beta(y+\varepsilon),\ \eta \in \beta_\varepsilon(y),\ \nu \in \beta(y-\varepsilon).$$

The set of operators satisfying (5.37), (5.38) is not empty as it contains β itself. Moreover, it is possible to construct a C^1 ε - uniform approximation β_ε. For this purpose, we consider a regularizing kernel, i.e. a C^∞ function $\rho: R \to R$ with support in $[0,1]$, positive and $\int_0^1 \rho(s)ds = 1$. The approximation is described in five basic cases, then we explain how to deal with the general situation.

Case 1. If $\mathrm{dom}(\beta) = R$, then $\beta_\varepsilon(s) = \int_0^1 \beta(s + \varepsilon\sigma)\rho(\sigma)d\sigma$. Here, it is important that β is single-valued a.e. and, obviously, β_ε satisfies all the desired properties.

Case 2. $\beta(s) \ni 0$ on $\mathrm{dom}(\beta) =]-\infty, s_0]$. We take

$$\beta_\varepsilon(s) = \begin{cases} 0 & \text{if } s < s_0, \\ \mathrm{tg}\frac{\pi}{2\varepsilon}(s - s_0) - \frac{\pi(s - s_0)}{2\varepsilon} & \text{if } s \in [s_0, s_0 + \varepsilon[, \\ \emptyset & \text{if } s \geq s_0 + \varepsilon, \end{cases}$$

which is maximal monotone and C^1. To check (5.37), (5.38) it is an easy task.

Case 3. $\beta(s) \ni 0$ on $\mathrm{dom}(\beta) = [s_0, +\infty[$. We take

$$\beta_\varepsilon(s) = \begin{cases} \emptyset & \text{if } s \leq s_0 - \varepsilon, \\ \mathrm{tg}\frac{\pi}{2\varepsilon}(s - s_0) - \frac{\pi(s - s_0)}{2\varepsilon} & \text{if } s \in]s_0 - \varepsilon, s_0], \\ 0 & \text{if } s > s_0. \end{cases}$$

Case 4. $\mathrm{dom}(\beta) =]- \ ,s_0[$. Then, necessarily $\beta(s) \to +\infty$ as $s \to s_0$, otherwise β would not be maximal. We define

$$\beta_\varepsilon(s) = \begin{cases} \int_0^1 \beta(s - \varepsilon\sigma)\rho(\sigma)d\sigma & \text{if } s \leq s_0 - \varepsilon \\ \int_0^1 \beta(s - \varepsilon\sigma)\rho(\sigma)d\sigma + \mathrm{tg}\frac{\pi}{2\varepsilon}(s - s_0 + \varepsilon) - \frac{\pi(s - s_0 + \varepsilon)}{2\varepsilon} & \text{if } s \in]s_0 - \varepsilon, s_0[, \\ \emptyset & \text{if } s \geq s_0. \end{cases}$$

Clearly, $\mathrm{dom}(\beta_\varepsilon) =]-\infty, s_0[$, β_ε is monotone and C^1 on $]-\infty, s_0[$. The condition (5.37) is an easy consequence of the definition.

Case 5. $\mathrm{dom}(\beta) =]s_0, +\infty[$. Similarly to case 4, we define

$$\beta_\varepsilon(s) = \begin{cases} \emptyset & \text{if } s \leq s_0, \\ \int_0^1 \beta(s + \varepsilon\sigma)\rho(\sigma)d\sigma + \mathrm{tg}\frac{\pi}{2\varepsilon}(s - s_0 - \varepsilon) - \frac{\pi(s - s_0 - \varepsilon)}{2\varepsilon} & \text{if } s_0 < s \leq s_0 + \varepsilon \\ \int_0^1 \beta(s + \varepsilon\sigma)\rho(\sigma)d\sigma & \text{if } s_0 + \varepsilon < s. \end{cases}$$

The general case may be obtained by the following property: if β^1 and β^2 are maximal monotone graphs in $R \times R$ as well as their sum $\beta = \beta^1 + \beta^2$ and β_ε^1, β_ε^2 are uniform approximations of β^1, β^2, then $\beta_\varepsilon^1 + \beta_\varepsilon^2$ is a uniform approximation of β. We also remark that, the following decomposition of maximal monotone graphs $\beta \subset R \times R$ is possible: if $s_1 \in \operatorname{intdom}(\beta)$ with $\beta(s_1)$ single-valued (we exclude the trivial case $\operatorname{dom}(\beta) = \{0\}$), we may write $\beta = \beta^1 + \beta^2$ with β^1, β^2 maximal monotone and $\operatorname{dom}(\beta^1) \supset]-\infty, s_1]$, $\operatorname{dom}(\beta^2) \supset [s_1, +\infty[$:

$$\beta^1(s) = \begin{cases} \beta(s_1) & s \le s_1, \\ \beta(s) & s \in \operatorname{dom}(\beta) \cap [s_1, +\infty[, \\ \emptyset & \text{otherwise}, \end{cases}$$

$$\beta^2(s) = \begin{cases} 0 & s \ge s_1, \\ \beta(s) - \beta(s_1) & s \in \operatorname{dom}(\beta) \cap]-\infty, s_1], \\ \emptyset & \text{otherwise}. \end{cases}$$

If $\operatorname{dom}(\beta^1)$ has the form $(-\infty, s_0[$, we may use case 4.
If $\operatorname{dom}(\beta^1)$ has the form $]-\infty, s_0]$, then we put

$$\beta^{1,a}(s) = \begin{cases} \beta^1(s) & \text{if } s < s_0, \\ \lim\limits_{s \to s_0} \beta^1(s) & \text{if } s \ge s_0, \end{cases}$$

$$\beta^{1,b}(s) = \begin{cases} 0 & \text{if } s < s_0, \\ [0, \infty[& \text{if } s = s_0, \\ \emptyset & \text{otherwise}. \end{cases}$$

Then $\beta^1 = \beta^{1,a} + \beta^{1,b}$ and the approximation is discussed in cases 1 and 2. By the decomposition property, we obtain the desired approximation of β^1 and, similarly of β^2.

To a C^1 uniform approximation β_ε we associate the perturbed equation

$$(5.39) \qquad \begin{aligned} &Ay_\varepsilon + \varphi(x, y_\varepsilon(x), u(x)) + \beta_\varepsilon(y_\varepsilon) = 0 && \text{in } \Omega, \\ &y_\varepsilon = 0 && \text{in } \partial\Omega. \end{aligned}$$

We are interested in feasible controls $u \in L^s(\Omega)$ $(s \ge 2)$ for the problem (5.27), in the sense that $u(x) \in K$ a.e. Ω and the mapping $[x,y] \to [\varphi(x,y,u(x)), L(x,y,u(x))]$ satisfies the conditions of Caratheodory, i.e. is continuous with respect to y a.e. $x \in \Omega$ and is measurable as a function of x for all y. These conditions imply that the mapping $x \to [\varphi(x,y(x),u(x)), L(x,y(x),u(x))]$ is measurable when $x \to y(x)$ is itself measurable.

<u>Theorem 5.15. Let</u> u <u>be feasible. Then</u> (5.28), (5.29) <u>and</u> (5.39), (5.29) <u>have unique solutions</u> $y, y_\varepsilon \in W^{2,s}(\Omega) \cap H_0^1(\Omega) \cap L^\infty(\Omega)$ <u>and</u> $|y_\varepsilon - y|_{L^\infty(\Omega)} \le \varepsilon$.

<u>Proof</u>

The existence, uniqueness and regularity of y, y_ε are standard results in the theory of elliptic variational inequalities. See Brezis [20], Kinderlehrer and Stampacchia [62] and Bonnans and Tiba [24] for a detailed treatment.

We prove that $y_\varepsilon(x) \leq y(x) + \varepsilon$ in (the converse inequality $y(x) \leq y_\varepsilon(x) + \varepsilon$ being proved similarly).

We define (we drop the variable $x \in \Omega$):

$$z = \min(0, y + \varepsilon - y_\varepsilon),$$

$$\mu_\varepsilon(y + \varepsilon) = \text{measurable selection of } \beta_\varepsilon(y + \varepsilon),$$

$$\delta = a(y,z) + \int_\Omega \varphi(\cdot, y + \varepsilon, u) z dx + \int_\Omega \mu_\varepsilon(y + \varepsilon) z dx .$$

As $z \in H_0^1(\Omega) \cap L^\infty(\Omega)$, the first two terms in δ are meaningful by (5.33), (5.34). For the third one, we remark that the product $\mu_\varepsilon(y(x) + \varepsilon) z(x)$ has to be understood to be given the value 0 if $z(x) = 0$. Otherwise $y(x) < y(x) + \varepsilon < y_\varepsilon(x)$, hence by (5.37) and the monotonicity of β_ε :

(5.40) $$\beta(y(x)) \leq \beta_\varepsilon(y(x) + \varepsilon) \leq \beta_\varepsilon(y_\varepsilon(x)) \qquad \text{a.e. } \Omega .$$

Then, defining

$$\mu(y(x)) = -(Ay(x) + \varphi(x, y(x), u(x))) \in L^s(\Omega),$$

$$\mu_\varepsilon(y_\varepsilon(x)) = -(Ay_\varepsilon(x) + \varphi(x, y_\varepsilon(x), u(x))) \in L^s(\Omega)$$

we have:

$$|\mu_\varepsilon(y(x) + \varepsilon) z(x)| \leq (|\mu(y(x))| + |\mu_\varepsilon(y_\varepsilon(x))|) |z(x)| .$$

The right hand side is in $L^s(\Omega)$ as $z \in L^\infty(\Omega)$, then δ is well defined. Extracting from δ the state equation, we get

$$\delta = \int_\Omega [Ay + \varphi(x,y,u) + \mu(y)] z dx + \int_\Omega [\varphi(x, y + \varepsilon, u) - \varphi(x,y,u)] \cdot z dx +$$
$$+ \int_\Omega [\mu_\varepsilon(y + \varepsilon) - \mu(y)] z dx .$$

Then, we notice that $\delta \leq 0$ since the first integral is null and $z \leq 0$.

We multiply (5.39) by z and we subtract from δ:

(5.41) $$\delta = a(y + \varepsilon - y_\varepsilon, z) + \int_\Omega [\varphi(\cdot, y + \varepsilon, u) - \varphi(\cdot, y_\varepsilon, u)] z dx + \int_\Omega [\mu_\varepsilon(y + \varepsilon) - \mu_\varepsilon(y_\varepsilon)] z dx$$

Here, we also use that $a(y + \varepsilon, z) = a(y,z)$.

By the monotonicity in y of $\varphi(\cdot,\cdot,\cdot)$ and of $\beta_\varepsilon(\cdot)$ and the definition of z, (5.41) yields

$$0 \geq a(y + \varepsilon - y_\varepsilon, z) = a(z,z) .$$

Since $z = 0$ on $\partial\Omega$, we deduce that $z = 0$ on Ω, hence $y_\varepsilon \leq y + \varepsilon$ a.e. Ω and the proof is finished.

Let us return now to the control problem (5.27) - (5.30). We approximate it by replacing (5.28) with (5.39).

<u>Theorem 5.16. For any feasible control</u> u, <u>the solutions</u> y_u^ε <u>of</u> (5.39) <u>and</u> y_u <u>of</u> (5.28) <u>satisfy</u>

(5.42) $$J(y_u^\varepsilon,u) = J(y_u,u) + O(\varepsilon)$$

<u>with</u> $|O(\varepsilon)| \leq C\cdot\varepsilon$. <u>If</u> K <u>is bounded</u>, C <u>is independent of</u> u.

<u>Proof</u>

We know that $|y_u^\varepsilon - y_u|_{L^\infty(\Omega)} \leq \varepsilon$. By (5.36) and a mean value theorem, we get

$$|J(y_u^\varepsilon,u) - J(y_u,u)| \leq \varepsilon \int_\Omega [M(x) + C|u(x)|]dx \cdot \eta(|y_u|_{L^\infty(\Omega)} + 1),$$

that is (5.42). If K is bounded, it follows that $|u|_{L^\infty(\Omega)}$ and $|y_u|_{L^\infty(\Omega)}$ are bounded, so we may take C independent of u.

<u>Remark 5.17.</u> If K is bounded, the relation(5.42) is also valid for the infimal values of the two problems.

We define the Hamiltonian

(5.43) $$H(x,y,u,p) = L(x,y,u) - p\varphi(x,y,u),$$

not to be confused with the similar notion from Ch. I, § 3 and Ch. II, § 1.

We use the space

$$E = \{u \in L^\infty(\Omega);\ u(x) \quad \text{is feasible}\}$$

with the Ekeland metric

$$d(u,v) = \text{meas}\{x \in \Omega\ ;\ u(x) \neq v(x)\}.$$

<u>Proposition 5.18.</u> (E,d) <u>is a complete metric space.</u>

<u>Proof</u>

Let $\{u_n\}$ be a Cauchy sequence in (E,d). Then, on a subsequence $\{u_{n_k}\}$, we may assume that

$$d(u_{n_k},u_{n_{k+1}}) \leq 2^{-k}.$$

We denote

$$A_k = \bigcup_{p \geq k} \{x \in \Omega\ ;\ u_{n_p}(x) \neq u_{n_{p+1}}(x)\}$$

and we notice that $\text{meas}(A_k) \leq \sum_{p=k}^{\infty} 2^{-p} = 2^{1-k}$ and that $A_k \supset A_{k+1}$. We define the function $\tilde{u}$ by

$$\tilde{u}(x) = u_{n_k}(x) \quad \text{if } x \notin A_k$$

and, obviously $u_{n_k} \to \tilde{u}$ in the metric d. Since the sequence is Cauchy, we have $u_n \to \tilde{u}$ in the metric d.

Since $u_n(x) \in K$ a.e. Ω, we get $\tilde{u}(x) \in K$ a.e. Ω. The mapping $[x,y] \to [\varphi(x,y,\tilde{u}(x)), L(x,y,\tilde{u}(x))]$ satisfies the Caratheodory conditions since the limit of a sequence of measurable functions is a measurable function. This shows that $\tilde{u}$ is feasible, i.e. $\tilde{u} \in E$, and the proof is finished.

Theorem 5.19. Assume that K is bounded. Let u be an α-solution of the problem (5.27)-(5.30) and put $\alpha_\varepsilon = \alpha + C\varepsilon$, where C is given by Thm. 5.16. There exists u_ε, α_ε-solution of the problem (5.27), (5.39), (5.29), (5.30), satisfying $d(u,u_\varepsilon) \leq (\alpha_\varepsilon)^{1/2}$ and such that, denoting y_ε the solution of (5.39) corresponding to u_ε and p_ε the costate associated to u_ε by

(5.44) $$A^* p_\varepsilon + \varphi'_y(\cdot,y_\varepsilon,u_\varepsilon)p_\varepsilon + \beta'_\varepsilon(y_\varepsilon)p_\varepsilon = L'_y(\cdot,y_\varepsilon,u_\varepsilon) \quad \text{in } \Omega$$

(5.45) $$p_\varepsilon = 0 \quad \text{in } \partial\Omega$$

with A^* the formal adjoint of A, the following relation is satisfied

(5.46) $$H(\cdot,y_\varepsilon,u_\varepsilon,p_\varepsilon) \leq H(\cdot,y_\varepsilon,v,p_\varepsilon) + (\alpha_\varepsilon)^{1/2} \quad \text{a.e. } \Omega$$

for any $v \in K$.

Remark. 5.20. By (5.33)-(5.36), we get $\varphi'_y(\cdot,y_\varepsilon,u_\varepsilon) \in L^s(\Omega)$, $L'_y(\cdot,y_\varepsilon,u_\varepsilon) \in L^s(\Omega)$, $\beta'_\varepsilon(y_\varepsilon) \in L^\infty(\Omega)$ since y_ε and u_ε belong to $L^\infty(\Omega)$. The existence of a weak solution $p_\varepsilon \in H^1_0(\Omega)$ for (5.44) may be easily obtained by a variational argument. Assuming that $L'_y(\cdot,y_\varepsilon,u_\varepsilon) \geq 0$ a.e. in Ω, the maximum principle gives $p_\varepsilon \geq 0$ a.e. in Ω. A comparison technique involving (5.44) and

$$A^* z = L'_y(\cdot,y_\varepsilon,u_\varepsilon) \quad \text{in } \Omega$$

$$z = 0 \quad \text{in } \partial\Omega$$

shows that $0 \leq p_\varepsilon(x) \leq z$ a.e. Ω, that is $p_\varepsilon \in L^\infty(\Omega)$. Coming back to (5.44), standard results for linear elliptic equations (Grisvard [54], Ch. II) give that $p_\varepsilon \in W^{2,s}(\Omega)$ too. Working with the positive and negative parts of $L'_y(\cdot,y_\varepsilon,u_\varepsilon)$, one may prove the regularity of p_ε in the general situation.

Proof of Thm. 5.19

By Thm. 5.16 u is an α_ε-solution of the problem (5.27), (5.39), (5.29), (5.30). The mapping $u \to y^\varepsilon_u$ given by (5.31) is continuous from (E,d) to $W^{2,s}(\Omega) \cap H^1_0(\Omega)$ endowed with the weak topology. To see this, let $d(u_k,u) \to 0$ and y_k be the solution of (5.39) corresponding to u_k. By the boundedness of K, we have $\{u_k\}$ bounded in $L^\infty(\Omega)$ and the standard estimates for variational inequalities give that $\{y_k\}$ is bounded in $W^{2,s}(\Omega) \cap H^1_0(\Omega)$. We may assume that $y_k \to y$ in the weak topology of $W^{2,s}(\Omega) \cap H^1_0(\Omega)$ and the Sobolev embedding theorem, Ch. I, §1, yields that $y_k \to y$ in $L^\infty(\Omega)$ strong. The Lebesgue theorem and (5.34) ive $\varphi(\cdot,y_k,u_k) \to \varphi(\cdot,y,u)$ in $L^s(\Omega)$. Obviously, $\beta_\varepsilon(y_k) \to \beta_\varepsilon(y)$ strongly in $L^\infty(\Omega)$ since β_ε is locally Lipschitzian.

Passing to the limit in (5.39) we obtain that $y_k \to y_u^\varepsilon$ weakly in $W^{2,s}(\Omega) \cap H_0^1(\Omega)$.

The conditions (5.35), (5.36) and the Lebesgue theorem yield that the mapping $u \to J(y_u^\varepsilon, u)$ is continuous from (E,d) to R.

Let $x_0 \in \Omega$ and $v \in K$. We say that $\{v_k\}$ is a sequence of variations of u around x_0 associated to v if

$$v_k = v \quad \text{in} \quad w_k(x_0) = \{x \in \Omega;\ |x - x_0| \leq 1/k\}$$

and $v_k(x) = u(x)$ otherwise. We denote $m_k(x_0) = \text{meas}(w_k(x_0))^{-1}$ and we have the following important lemma

Lemma 5.21. *Let u and v be as above. Then*

$$\lim_{k \to \infty} m_k(x_0)[J(y_k^\varepsilon, v_k) - J(y_u^\varepsilon, u)] = H(x_0, y_u^\varepsilon(x_0), v, p_u^\varepsilon(x_0)) - \tag{5.47}$$
$$- H(x_0, y_u^\varepsilon(x_0), u(x_0), p_u^\varepsilon(x_0)) \qquad \text{a.e. } x_0 \in \Omega,$$

where y_k^ε is the solution of (5.39) associated to v_k.

Proof

We have

$$J(y_k^\varepsilon, v_k) - J(y_u^\varepsilon, u) = \quad [L(\cdot, y_k^\varepsilon, v_k) - L(\cdot, y_u^\varepsilon, u)]dx =$$

$$= \int_\Omega [L(\cdot, y_u^\varepsilon, v_k) - L(\cdot, y_u^\varepsilon, u)]dx + \int_\Omega [L(\cdot, y_k^\varepsilon, v_k) - L(\cdot, y_u^\varepsilon, v_k)]dx$$

The last integral may be rewritten as

$$\int_\Omega L_y'(\cdot, \tilde{y}_k, v_k)(y_k^\varepsilon - y_u^\varepsilon)dx = \int_\Omega [A^*\tilde{p} + \varphi_y'(\cdot, \hat{y}_k, v_k)\tilde{p}_k + \beta_\varepsilon'(\ddot{y}_k)\tilde{p}_k] \cdot$$
$$\cdot (y_k^\varepsilon - y_u^\varepsilon)dx = \int_\Omega A(y_k^\varepsilon - y_u^\varepsilon)\tilde{p}_k dx + \int_\Omega \varphi_y'(\cdot, \hat{y}_k, v_k)(y_k^\varepsilon - y_u^\varepsilon)\tilde{p}_k dx +$$
$$+ \int_\Omega \beta_\varepsilon'(\ddot{y}_k)\tilde{p}_k(y_k^\varepsilon - y_u^\varepsilon)dx = \int_\Omega [\varphi(\cdot, y_u^\varepsilon, u) - \varphi(\cdot, y_k^\varepsilon, v_k) + \beta_\varepsilon(y_u^\varepsilon) - \beta_\varepsilon(y_k^\varepsilon)] \cdot \tilde{p}_k dx +$$
$$+ \int_\Omega [\varphi(\cdot, y_k^\varepsilon, v_k) - \varphi(\cdot, y_u^\varepsilon, v_k) + \beta_\varepsilon(y_k^\varepsilon) - \beta_\varepsilon(y_u^\varepsilon)]\tilde{p}_k =$$
$$= \int_\Omega [\varphi(\cdot, y_u^\varepsilon, u) - \varphi(\cdot, y_u^\varepsilon, v_k)]\tilde{p}_k dx,$$

where $\tilde{y}_k$, $\hat{y}_k$, $\ddot{y}_k$ are intermediay points given by the mean value theorem applied to $L, \varphi, \beta_\varepsilon$ respectively and $\tilde{p}_k \in W^{2,s}(\Omega) \cap H_0^1(\Omega)$ satisfies (see Remark 5.20):

$$A^*\tilde{p}_k + \varphi_y'(\cdot, \hat{y}_k, v_k)\tilde{p}_k + \beta_\varepsilon'(\ddot{y}_k)\tilde{p}_k = L_y'(\cdot, \tilde{y}_k, v_k) \qquad \text{in } \Omega \tag{5.48}$$

(5.49) $\qquad \tilde{p}_k = 0 \qquad\qquad$ in $\partial\Omega$.

Then, we have:

(5.50)
$$J(y^\varepsilon_k, v_k) - J(y^\varepsilon_u, u) = \int_\Omega [L(\cdot, y^\varepsilon_u, v_k) - \varphi(\cdot, y^\varepsilon_u, v_k)\tilde{p}_k - L(\cdot, y^\varepsilon_u, u) + \varphi(\cdot, y^\varepsilon_u, u)\tilde{p}_k]dx =$$
$$= \int_{w_k(x_0)} [H(\cdot, y^\varepsilon_u, v, \tilde{p}_k) - H(\cdot, y^\varepsilon_u, u, \tilde{p}_k)]dx \; .$$

We notice that $\{v_k\}$ is bounded in $L^\infty(\Omega)$ and $v_k \to u$ a.e. Ω . Then, standard estimates for variational inequalities show that $y^\varepsilon_k \to y^\varepsilon_u$ weakly in $W^{2,s}(\Omega) \cap H^1_o(\Omega)$ (ε is fixed here). Consequently, $\hat{y}_k$, $\ddot{y}_k$, $\tilde{y}_k \to y^\varepsilon_u$ uniformly in Ω and, under conditions (5.33)-(5.36), $\varphi'_y(\cdot, \hat{y}_k, v_k) \to \varphi'_y(\cdot, y^\varepsilon_u, u)$, $\beta'_\varepsilon(\ddot{y}_k) \to \beta'_\varepsilon(y^\varepsilon_u)$, $L'_y(\cdot, \tilde{y}_k, v_k) \to L'_y(\cdot\, y^\varepsilon_u, u)$ at least in $L^s(\Omega)$.

As in Remark 5.20, it is standard to obtain the boundedness of $\{\tilde{p}_k\}$ in $W^{2,s}(\Omega) \cap H^1_o(\Omega)$. The above convergences and the Sobolev imbedding theorem show that $\tilde{p}_k \to p^\varepsilon_u$ (the solution of (5.44) associated to u and y^ε_u), uniformly in Ω .

We use (5.50) to obtain

(5.51)
$$J(y^\varepsilon_k, v_k) - J(y^\varepsilon_u, u) = \int_{w_k(x_0)} [H(\cdot, y^\varepsilon_u, v, \tilde{p}_k) - H(\cdot, y^\varepsilon_u, u, \tilde{p}_k)]dx =$$
$$= \int_{w_k(x_0)} [H(\cdot, y^\varepsilon_u, v, p^\varepsilon_u) - H(\cdot, y^\varepsilon_u, u, p^\varepsilon_u)]dx + \int_{w_k(x_0)} \varphi(\cdot, y^\varepsilon_u, u)(p^\varepsilon_u - \tilde{p}_k)dx +$$
$$+ \int_{w_k(x_0)} \varphi(\cdot, y^\varepsilon_u, u)(\tilde{p}_k - p^\varepsilon_u)dx \, .$$

Multiplying by $m_k(x_0)$ and taking into account that a.e. $x_0 \in \Omega$ is a Lebesgue point for an integrable application and the uniform convergence $\tilde{p}_k - p^\varepsilon_u \to 0$, we get (5.47).

Proof of Theorem 5.19 (continued)

We apply the Ekeland variational principle to u and the problem (5.27), (5.39), (5.29), (5.30). This ensures the existence for this problem of an α_ε - solution denoted u_ε , satisfying $d(u, u_\varepsilon) \le (\alpha_\varepsilon)^{1/2}$ and

(5.52) $\qquad J(y_\varepsilon, u_\varepsilon) \le J(y^\varepsilon_z, z) + (\alpha_\varepsilon)^{1/2} d(u_\varepsilon, z), \qquad \forall\, z \in E,$

where y^ε_z is the solution of (5.39) corresponding to z and y_ε is defined in the statement of Thm. 5.19.

We take z to be given by a sequence v_k of variations of u_ε around $x_0 \in \Omega$, associated to $v \in K$ and divide (5.52) by $d(u_\varepsilon, v_k) = meas(w_k(x_0))$. Then, Lemma 5.21 gives (5.46) and finishes the proof.

Theorem 5.22. We assume that K is bounded. Let $\bar{u}$ be a solution of the original problem and $\bar{y}$ the corresponding solution of (5.28) and u_ε be given by Thm. 5.19 applied to $\bar{u}$. Then, for $\varepsilon \to 0$, we have

(5.53) $d(u_\varepsilon, \bar{u}) \to 0,$

(5.54) $y_\varepsilon \to \bar{y}$ <u>weakly in</u> $W^{2,s}(\Omega) \cap H_0^1(\Omega)$

<u>and there exists</u> $\bar{p}$, <u>limit point in</u> $H_0^1(\Omega)$ <u>weak and</u> $L^\infty(\Omega)$ <u>weak star of</u> p_ε <u>and</u> ρ , <u>limit point in</u> $H^{-1}(\Omega)$ <u>weak of</u> $\beta_\varepsilon'(y_\varepsilon)p_\varepsilon$ <u>such that</u>

(5.55) $A^*\bar{p} + \varphi_y'(\cdot, \bar{y}, \bar{u})\bar{p} + \rho = L_y'(\cdot, \bar{y}, \bar{u})$ <u>in</u> Ω,

(5.56) $\bar{p} = 0$ <u>in</u> $\partial\Omega$

<u>and for all</u> $v \in K$

(5.57) $H(\cdot, \bar{y}, \bar{u}, \bar{p}) \leq H(\cdot, \bar{y}, v, \bar{p})$ a.e. Ω .

Proof

By <u>Thm. 5.19</u>, we get (5.53) and (5.54) is a consequence of standard estimates for variational inequalities.

As in the proof of <u>Lemma 5.21</u> and in <u>Remark 5.20</u>, we obtain that $\{p_\varepsilon\}$ is bounded in $H_0^1(\Omega) \cap L^\infty(\Omega)$ and (5.44) gives that $\{\beta_\varepsilon'(y_\varepsilon)p_\varepsilon\}$ is bounded in $H^{-1}(\Omega)$. We may pass to the limit in (5.44) (on a subsequence) in the weak topology of $H^{-1}(\Omega)$ and prove (5.55), (5.56).

As $H_0^1(\Omega)$ is compactly imbedded in $L^2(\Omega)$, then, on a subsequence, $p_\varepsilon \to \bar{p}$ a.e. in Ω and we get that $H(\cdot, y_\varepsilon, u_\varepsilon, p_\varepsilon) \to H(\cdot, \bar{y}, \bar{u}, \bar{p})$ and $H(\cdot, y_\varepsilon, v, p_\varepsilon) \to H(\cdot, \bar{y}, v, \bar{p})$ a.e. in Ω . Since $\alpha_\varepsilon \to 0$, (5.46) implies (5.57) and ends the proof.

In order to obtain more information on ρ , we assume as in §1, §2, this chapter, that β is the maximal monotone extension of a monotone simple function defined on some real interval. That is, the graph of β is composed only of segments parallel to the axes. We denote

$D = \{ r \in R;\ r$ is a discontinuity point of $\beta \}$,

$\Omega_o = \{ x \in \Omega ;\ \bar{y}(x) \notin D \}$,

which is an open subset in Ω because $\bar{y}$ is continuous.

Corollary 5.23. Under the above assumptions, the distribution ρ <u>satisfies</u> supp $\rho \subset \Omega - \Omega_o$.

Proof

Let $\Psi \in \mathcal{D}(\Omega)$ with supp $\Psi \subset \Omega_o$, and $c > 0$ be such that

$$\inf_{x \in \text{supp}\Psi} \text{dist}(\bar{y}(x), D) \geq 2c.$$

It follows that

$$\inf_{x \in \text{supp}\Psi} \text{dist}(y_\varepsilon(x), D) \geq c > 0 .$$

by the uniform convergence of y_ε and for ε sufficiently small.

Condition (5.37) shows that $\beta_\varepsilon(r)$ is constant when $\mathrm{dist}(r,D) \geq c$ and ε is small. Then $\beta_\varepsilon'(y_\varepsilon) = 0$ for $x \in \mathrm{supp}\Psi$ and ε small. It yields that

$$\rho(\Psi) = \lim_{\varepsilon \to 0} \int_\Omega \beta_\varepsilon'(y_\varepsilon) p_\varepsilon \Psi \, dx = 0,$$

which finishes the proof.

Remark 5.24. The results of this section are based mainly on the works of Bonnans and Casas [23], Bonnans and Tiba [24].

Remark 5.25. Starting with the pionnering work of Mignot [74], much attention is paid in the literature to the control of elliptic variational inequalities. We quote the book of Barbu [13] for a detailed treatment by the adapted penalization method and for more references. Recently, Shi Shuzhong [105] studied an abstract variant, but his proof contains a gap and the result is not correct. In this setting, the works of Barbu and Tiba [19] and Wenbin and Rubio [137] discuss the abstract problem and propose some variants of the maximum principle.

IV. FREE BOUNDARY PROBLEMS

One of the main motivations of the interest for the theory of variational inequalities is given by their intimate relationship with the free boundary problems. These are partial differential equations with boundary and initial conditions, on a domain with an unknown moving part of the boundary and having to be found together with the solution.

Among the most studied free boundary problems we quote the dam problem and the Stefan problems (Friedman [47], Rubenstein [100], Ockendon and Elliott [45]). By the Baiocchi [25] transform the dam problem is reduced to an elliptic variational inequality and the one phase Stefan problem becomes a parabolic variational inequality. By this approach, they are discussed from the point of view of the control problems in the book by V.Barbu [13]. The results from the previous chapter may be also interpreted in this sense. Moreover, in the recent works [48], [49], [50], [51], situations of this type are investigated when a complete determination of the optimal control is possible.

We devote this chapter exclusively to the study of the optimal control problems governed by two-phase Stefan problems. Our choice is also motivated by the special interest for applications of this type of problem, Rubenstein [101], Saguez [109], Bonnans [22].

Since the two-phase Stefan problems form the object of recent research, we start with a section on this subject.

1. Two-phase Stefan problems

The Stefan problems are models of heat propagation processes with change of phase, that is with melting or solidification. The simplest examples are the melting of ice in water or the melting and the solidification of metals in metallurgy. From this point of view, one may see the importance for applications of Stefan problems.

The separation surface between the two phases, which is unknown and has an evolution in time, is called the free boundary. If we assume that $S(x,t) = 0$, $x \in R^N$, $t \in [0,T]$, is a representation of it, then the conservation of the energy at the free boundary is expressed by the law of J.Stefan [107]:

$$(1.1) \qquad [K \operatorname{grad} u . \operatorname{grad} S]_{solid}^{liquid} = \rho L \partial S / \partial t,$$

where K is the thermal conductivity, u is the temperature, ρ is the density and L is the specific latent heat.

When the temperature u is assumed to vary in one phase only, while in the other is preserved constant, we obtain the one-phase Stefan problem (§1, Ch.III, Example 3). When the temperature u may vary in both phases, we have two-phase Stefan problems, which form the object of this section.

Suppose that the solidification/melting temperature is $u = 0$. Let Ω be a bounded domain in R^N with regular boundary Γ, $Q =]0,T[\times\Omega$ and $Q_1 = \{(t,x)\in Q; u(t,x) < 0\}$, $Q_2 = \{(t,x)\in Q; u(t,x) > 0\}$ and $S(t)$ be the separating surface between the two phases at the moment $t\in[0,T]$.

In what follows, we limit ourselves to the following simplified model:

$$(1.2) \qquad c_i \partial u/\partial t - K_i \Delta u = g \qquad \text{in } Q_i,\ i=1,2,$$

where $c_i > 0$ is the specific heat in Q_i and $K_i > 0$ is the thermal conductivity in Q_i. We remark that, generally, c_i and K_i may depend on the temperature u, but by the Kirchoff transform ([36], p.45, [74]) one may obtain the diffusion equation (1.2), with constant coefficients in Q_i.

On the free boundary $S(t)$, we have $u(t,x) = 0$ and the Stefan condition (1.1) becomes (we take $\rho = 1$):

$$(1.3) \qquad K_1 \operatorname{grad} u_1 \cdot n - K_2 \operatorname{grad} u_2 . n = LV.n,$$

where $u_i = u|_{Q_i}$, $i = 1,2$, n is the normal to $S(t)$ and V is the propagation speed of $S(t)$.

On the fixed boundary of Q, we impose Dirichlet or Neumann conditions and we also choose an initial condition $u = u_0$.

The problem (1.2), (1.3) with the mixed conditions given above, may be transformed into a problem on the fixed domain Q, by the enthalpy method.

Let $\beta\subset R \times R$ be the maximal monotone graph

$$(1.4) \qquad \beta(u) = \begin{cases} c_1 u & u < 0, \\ [0,L] & u = 0, \\ c_2 u + L & u > 0. \end{cases}$$

The equation (1.2) may be rewritten as

$$(1.5) \qquad \partial y/\partial t - \Delta u = g \qquad \text{in } Q_1 \cup Q_2,$$

$$(1.6) \qquad y\in\beta(u) \qquad \text{in } Q_1 \cup Q_2.$$

We have fixed $K_i = 1$, $i=1,2$, which may be easily obtained by dividing (1.2) by $K_i > 0$ and redenoting c_i.

If $S\in C^1(Q)$ and u and its derivatives from (1.2) are continuous, satisfying (1.2), (1.3) and the mixed conditions, then the pair u, S is called a classical solution of the Stefan problem. In the case of many spatial variables, the existence of local classical solutions is proved in A. Meirmanov [78]. In space dimension 1, the existence of global classical solutions is known [100].

In the papers of Kamenomotskaya [64], [65] a weak solution is defined for (1.5), (1.6), (1.3) and Dirichlet conditions $u|_\Sigma = v$, by

$$(1.7)\qquad \int_Q (y\partial\phi/\partial t + u\Delta\phi) = -\int_\Omega y_0\phi(x,0) + \int_\Sigma v\partial\phi/\partial n - \int_Q g\phi,$$

where y_0 is the initial value of the enthalpy and ϕ is a test function from

$$(1.8)\qquad \mathcal{F} = \{\phi\in C^1(Q);\ \partial^2\phi/\partial x_i\partial x_j \in C(\bar{Q}),\ \phi(x,T)=0,\ \phi(x,t)=0 \text{ on } \Sigma\}.$$

Multiplying by ϕ in (1.5), integrating by parts and using (1.2), we see that any classical solution is a weak solution too. The converse is valid under certain regularity assumptions on u,y, [45], p.64.

If the data v, u_0 are sufficiently regular and satisfy some compatibility conditions, then the Stefan problem has a unique weak solution $u\in H^1(Q)$, $y\in L^\infty(0,T;L^2(\Omega))$, [45], [65]. In this setting, the free boundary is obtained as the set of points $(t,x)\in Q$, such that $u(t,x) = 0$.

Regularity properties for the weak solution, including global continuity, have been recently obtained by Caffarelli and Evans [37], E.DiBenedetto [28].

Remark 1.1. It doesn't follow necessarily that the set where $u = 0$ has zero measure in Q. This fact was remarked by Atthey [8] and, as a consequence, three-phase models were considered. The third phase is characterized by constant temperature $u = 0$ and is called the "mushy region". A classical formulation of three-phase Stefan problems and a study of the regularity properties of the weak solution and of the boundaries of the mushy region, are due to Primicerio [89], Ughi [135].

By (1.7), in particular, it yields that we have

$$(1.9)\qquad \partial y/\partial t - \Delta u = g \qquad \text{in } \mathcal{D}'(Q).$$

We remark that β^{-1} is a continuous, monotone mapping on R and (1.9) may be rewritten as

$$(1.10)\qquad \partial y/\partial t - \Delta\beta^{-1}(y) = g \qquad \text{in } \mathcal{D}'(Q),$$

which suggests the semigroup approach to the two-phase Stefan problems, Evans [37], Barbu [10], p.205.

We apply in (1.9) the transformation

$$(1.11)\qquad w(t,x) = \int_0^t u(s,x)ds$$

due to Duvaut [41], Fremond [53]. We obtain the problem

$$(1.12)\qquad \beta(\partial w/\partial t) - \Delta w \ni f \qquad \text{in } Q,$$

$$(1.13)\qquad w(0,x) = 0 \qquad \text{in } \Omega,$$

with the corresponding boundary conditions. Here, we have

(4.14) $\qquad f(t,x) = \int_0^t g(s,x)ds + y_0(x)$ in Q.

The problem (1.12), (1.13) is a second kind parabolic variational inequality and existence as well as uniqueness results for the solution $w \in W^{1,2}(0,T;L^2(\Omega))$ may be found in the monograph of V.Barbu [12], Ch.IV, § 3. We call $u = \partial w/\partial t$ the variational solution of the Stefan problem.

Obviously any weak solution is a variational solution too. If the existence and uniqueness of both types of solutions is fulfilled, then the converse is also valid.

Remark 1.2. For other types of solutions, interesting in applications, we quote Ockendon and Elliott [45], Ch.III. An important number of recent papers discuss generalized Stefan problems, including supplementary nonlinear terms in the equation or in the boundary conditions, sources on the free boundary or the degeneracy of the equation ($\beta = 0$ in certain regions). In this respect we mention the papers [82], [83], [85], [92], [111] and their references.

Another generalization of interest is related to the melting or solidification of alloys, for which the change of phase temperature is unknown and depends on the concentration of the different components. The mathematical formulation of these problems is based on the theory of quasi-variational inequalities and a theoretical and numerical study may be found in the work of Saguez and Bermudez [108].

As concerns the rich literature on the numerical treatment of two-phase Stefan problems, we shall make some comments in section 4.

We close these considerations with two specific existence results which are needed in the sequel. We remark that β given by (1.4) is a maximal monotone graph in R x R, satisfying

(1.14) $\qquad (\beta(y) - \beta(z))(y - z) \geq \alpha(y - z)^2, \alpha > 0$

and, for the sake of simplicity, we take $\alpha = 1$.

We begin with the case of Dirichlet boundary conditions. Let $H = L^2(\Omega)$ and θ, π be the operators defined in $L^2(0,T;H)$ by $\theta g = u$, $\pi g = y$, where u is the variational solution of (1.9) with the mixed conditions

(1.15) $\qquad y(0) = y_0 \qquad$ in Ω,

(1.16) $\qquad u(t,x) = 0 \qquad$ in Σ.

Proposition 1.3. We assume (1.14) and $y_0 \in H$, $u_0 = \beta^{-1}(y_0) \in H_0^1(\Omega)$. Then θ, π are defined on all $L^2(0,T;H)$, are weakly-strongly continuous from $L^2(0,T;H)$ to $C(0,T;H)$, $C(0,T;H^{-1}(\Omega))$ respectively and, moreover, we have the estimates:

(1.17) $\quad |(\theta g)_t|_{L^2(0,T;H)} + |\theta g|_{L^\infty(0,T;H_0^1(\Omega))} \leq C(1 + |g|_{L^2(0,T;H)})$,

(1.18) $\quad |(\pi g)_t|_{L^2(0,T;H^{-1}(\Omega))} + |\pi g|_{L^\infty(0,T;H)} \leq C(1 + |g|_{L^2(0,T;H)})$,

for all $g \in L^2(0,T;H)$, where $C = C(|y_0|_H, |u_0|_{H_0^1(\Omega)})$.

Proof

Let $A : H \to H$, $\mathrm{dom}(A) = H_0^1(\Omega) \cap H^2(\Omega)$, $Au = -\Delta u$. This may be viewed as a maximal monotone operator, even a subdifferential, according to Example 4, Ch. I, § 3.1. Let A_λ be the Yosida approximation of A. We consider the approximating problem

(1.19) $\quad \beta(u_\lambda)_t + A_\lambda u_\lambda \ni g \quad$ in $]0,T[\times\Omega$,

(1.20) $\quad \beta(u_\lambda(0)) \ni y_0$.

Equivalently, (1.19), (1.20) may be rewritten

(1.21) $\quad u_\lambda(t) = \beta^{-1}\left\{ y_0 + \int_0^t [g(s) - A_\lambda u_\lambda(s)]ds \right\}.$

Since, by (1.14), β^{-1} is Lipschitzian and A_λ is Lipschitzian too, the equation (1.21) has a unique solution $u_\lambda \in W^{1,2}(0,T;H)$.

Let $y_\lambda \in W^{1,2}(0,T;H)$ be given by

(1.22) $\quad y_\lambda(t) = y_0 + \int_0^t [g(s) - A_\lambda u_\lambda(s)]ds \in \beta(u_\lambda(t)).$

We multiply (1.19) by $(u_\lambda)_t$, respectively y_λ, and a standard argument, based on (1.14), implies

(1.23) $\quad |(u_\lambda)_t|_{L^2(0,T;H)} + |(I+\lambda A)^{-1}u_\lambda|_{L^\infty(0,T;H_0^1(\Omega))} \le C(1 + |g|_{L^2(0,T;H)})$,

(1.24) $\quad |(y_\lambda)_t|_{L^2(0,T;H^{-1}(\Omega))} + |y_\lambda|_{L^\infty(0,T;H)} \le C(1 + |g|_{L^2(0,T;H)})$.

We also obtain $\left\{ \lambda^{1/2} A_\lambda u_\lambda \right\}$ bounded in $L^\infty(0,T;H)$ and the Arzela-Ascoli theorem gives

$(I + \lambda A)^{-1}u_\lambda \to u$ strongly in $C(0,T;H)$,
weakly* in $L^\infty(0,T;H_0^1(\Omega))$,
$u_\lambda \to u$ strongly in $C(0,T;H)$,
$(u_\lambda)_t \to u_t$ weakly in $L^2(0,T;H)$,
$y_\lambda \to y \in \beta(u)$ strongly in $C(0,T;H^{-1}(\Omega))$,
weakly* in $L^\infty(0,T;H)$,
$(y_\lambda)_t \to y_t$ weakly in $L^2(0,T;H^{-1}(\Omega))$.

Passing to the limit in (1.19), (1.20) we see that u is the solution of the problem (1.9), (1.15), (1.16), so $\mathrm{dom}\,\theta = \mathrm{dom}\,\pi = L^2(0,T;H)$ and the estimates (1.17), (1.18) are fulfilled. The weak-strong continuity of θ, π is an easy consequence of these estimates.

To prove the uniqueness of the solution, let u, v be two solutions of the problem, so

(1.25) $\quad \beta(u)_t - \beta(v)_t - \Delta(u - v) \ni 0.$

We multiply (1.25) by $\beta(u) - \beta(v)$ in the scalar product of $H^{-1}(\Omega)$. Using (1.14), we deduce

$$1/2\, d/dt\, |\beta(u)-\beta(v)|^2_{H^{-1}(\Omega)} + |u(t)-v(t)|^2_H \le 0,$$

that is u = v. This ends the proof.

Now, we turn to the case of Neumann boundary conditions

(1.26) $$\partial u/\partial n = e \qquad \text{on } \Sigma,\ \ e \in L^2(\Sigma).$$

Let $\tilde{A}: H^1(\Omega) \to H^1(\Omega)^*$ be the linear, continuous symmetric operator defined by

(1.27) $$(\tilde{A}u,v) = \int_\Omega \text{grad } u.\text{grad } v dx,\ u,v \in H^1(\Omega).$$

The problem (1.9), (1.15), (1.26) may written as an evolution equation in abstract spaces:

(1.28) $$dy/dt + \tilde{A}u = g_1,$$

(1.29) $$y \in \tilde{\beta}(u),$$

(1.30) $$y(0) = y_o,$$

where $g_1 \in L^2(0,T;H^1(\Omega)^*)$ is defined by

(1.31) $$\int_0^T (g_1(t),\Psi(t))_{H^1(\Omega)\times H^1(\Omega)^*} = \int_0^T\int_\Gamma e.\Psi + \int_Q g.\Psi$$

for any $\Psi \in L^2(0,T;H^1(\Omega))$, and $\tilde{\beta}$ is the realization of β in $L^2(\Omega) = H$.

We apply the transformation (1.11) and obtain the equation

(1.32) $$\tilde{\beta}(dw/dt) + \tilde{A}w \ni f \qquad \text{in } [0,T],$$

(1.33) $$w(0) = 0,$$

where $f(t) = \int_0^t g_1(s)ds + y_o$, $f \in W^{1,2}(0,T;H^1(\Omega)^*)$.

<u>Proposition 1.4. The problem</u> (1.28) - (1.30) <u>has a unique variational solution</u> $u \in L^2(0,T;H)$, $y \in L^2(0,T;H)$. <u>In addition</u> $w \in C(0,T;H^1(\Omega))$.

<u>Proof</u>

We consider the approximating equation

(1.34) $$\tilde{\beta}(dw_\varepsilon/dt) + \varepsilon w_\varepsilon + \tilde{A}w_\varepsilon \ni f,\ \varepsilon > 0.$$

As $\varepsilon I + \tilde{A}$ is coercive on $H^1(\Omega)$, it yields that (1.33), (1.34) has a unique solution $w_\varepsilon \in C(0,T;H^1(\Omega))$, $dw_\varepsilon/dt \in L^2(0,T;H)$, $\sqrt{t}\, dw_\varepsilon/dt \in L^2(0,T;H^1(\Omega))$, according to Barbu [12], p.213.

We multiply by dw_ε/dt and integrate over [0,t]

$$\int_0^t |dw_\varepsilon/dt|^2_H + 1/2(\tilde{A}w_\varepsilon(t), w_\varepsilon(t)) \le \int_0^t (f, dw_\varepsilon/dt).$$

Integrating by parts on [0,t] in the right hand side, we get $\{w_\varepsilon\}$ bounded in $L^\infty(0,T;H^1(\Omega))$, $\{dw_\varepsilon/dt\}$ bounded in $L^2(0,T;H)$. By (1.4), we see that $\{\tilde{\beta}(dw_\varepsilon/dt)\}$ is bounded in $L^2(0,T;H)$.

We subtract two equations (1.34), corresponding to the parameters $\varepsilon,\lambda > 0$ and we multiply by $dw_\varepsilon/dt - dw_\lambda/dt$. A simple reasonning shows that

$$dw_\varepsilon/dt \to dw/dt \qquad \text{strongly in } L^2(0,T;H),$$
$$w_\varepsilon \to w \qquad \text{strongly in } C(0,T;H^1(\Omega)),$$

and, by the demiclosedness of maximal monotone operators, we have

$$\tilde{\beta}(dw_\varepsilon/dt) \to \tilde{\beta}(dw/dt) \qquad \text{weakly in } L^2(0,T;H).$$

Remark 1.5. In section 3, Proposition 3.1 we show that the variational solution of (1.28) - (1.30) satisfies $u \in L^2(0,T;H^1(\Omega))$. The result given above appears as a particular case in the paper of E.DiBenedetto and R.E.Showalter [27], where A may be nonlinear too. The direct argument, indicated here, will be used in § 3 to prove the continuous dependence on the right hand side of the solution, in the weak topology.

Remark 1.6. Since we use the form (1.9) and not the form (1.10) of the Stefan problem, we cannot apply the abstract scheme from § 1, Chapter II, directly. However, we show that the conditions (a), (b), (c) continue to play an important role.

Remark 1.7. If $\beta(r) = r^{1/m}\,\mathrm{sign}\, r$, $m > 1$, then the equation (1.9) models the diffusion of gases in a porous medium, W.F.Ames [4]. Regularity results for this situation were obtained by L.A.Caffarelli and A.Friedman [38], E.DiBenedetto [28].

2. Distributed control

We consider the problem

(2.1) $\quad$ Minimize $\int_0^T (1/2\,|y - y_d|_H^2 + \Psi(u))dt$,

(2.2) $\quad v_t - \Delta y = Bu + f \qquad$ in Q,

(2.3) $\quad v \in \beta(y) \qquad$ in Q,

(2.4) $\quad v(0,x) = v_0(x) \qquad$ in Ω,

(2.5) $\quad y(t,x) = 0 \qquad$ in Σ.

Here U is a control Hilbert space, $u \in L^2(0,T;U)$, y is the state of the system and v is the enthalpy, while $\beta \subset R \times R$ is a maximal monotone graph of the form (1.4). The operator $B : U \to H$ is linear, continuous, $f, y_d \in L^2(0,T;H)$, $v_0 \in H$, $y_0 = \beta^{-1}(v_0) \in V = H_0^1(\Omega)$ and $\Psi : U \to]-\infty,+\infty]$ is convex, lower semicontinuous, proper. In particular, if we have control constraints $u \in U_{ad}$, a closed, convex subset of U, these may be implicitly expressed by redefining $\Psi = +\infty$ outside U_{ad}.

If Ψ is coercive, the Proposition 1.3 implies the existence of at least one optimal pair $[y^*,u^*]$ for the problem (2.1) - (2.5).

By (1.14), we see that $\gamma=\beta-I$ (I is the identity of R) is maximal monotone in R x R. We regularize β by

$$\beta^{\varepsilon}(y) = y + \int_{-1}^{1} \gamma_{\varepsilon}(y - \varepsilon^2\tau)\rho(\tau)d\tau = y + \gamma^{\varepsilon}(y), \tag{2.6}$$

where ρ is a Friedrichs mollifier and γ_{ε} is the Yosida approximation of γ.

We denote by $\theta_{\varepsilon}, \pi_{\varepsilon}$ the operators obtained as θ, π from (1.9), (1.15), (1.16), when β is replaced by β^{ε} from (2.6).

Lemma 2.1. Let $g, g_{\varepsilon} \in L^2(0,T;H)$, $g_{\varepsilon} \to g$ weakly in $L^2(0,T;H)$. Then:

$$\theta_{\varepsilon}(g_{\varepsilon}) \to \theta(g) \quad \text{strongly in } C(0,T;H), \tag{2.7}$$

$$\theta_{\varepsilon}(g_{\varepsilon})_t \to \theta(g)_t \quad \text{weakly in } L^2(0,T;H), \tag{2.8}$$

$$\pi_{\varepsilon}(g_{\varepsilon}) \to v \in \beta(\theta(g)) \quad \text{strongly in } C(0,T;V^*) \tag{2.9}$$

$$\pi_{\varepsilon}(g_{\varepsilon})_t \to v_t \quad \text{weakly in } L^2(0,T;V^*). \tag{2.10}$$

The argument is the same as in Proposition 1.3.

For each $\varepsilon > 0$, we denote $\psi_{\varepsilon}: U \to]-\infty, +\infty]$ the regularization of ψ

$$\psi_{\varepsilon}(h) = \inf\left\{ |h - v|_U^2/2\varepsilon + \psi(v),\ v \in U \right\}. \tag{2.11}$$

We consider the approximating control problem:

$$\text{Minimize} \int_0^T (1/2|y - y_d|_H^2 + 1/2|u - u^*|_U^2 + \psi_{\varepsilon}(u))dt, \tag{2.12}$$

for $u \in L^2(0,T;U)$, $y = \theta_{\varepsilon}(u) \in W^{1,2}(0,T;H)$. The cost functional (2.12) is coercive in u (uniformly with respect to ε) and we deduce the existence of the optimal pairs $[y_{\varepsilon}, u_{\varepsilon}]$ for the problem (2.12).

Lemma 2.2. When $\varepsilon \to 0$, we have

$$y_{\varepsilon} \to y^* \quad \text{strongly in } C(0,T;H), \tag{2.13}$$

$$u_{\varepsilon} \to u^* \quad \text{strongly in } L^2(0,T;U). \tag{2.14}$$

Proof

Let J_{ε} denote the functional (2.12). Then

$$J_{\varepsilon}(y_{\varepsilon}, u_{\varepsilon}) \le \int_0^T (1/2|\theta_{\varepsilon}(Bu^* + f) - y_d|_H^2 + \psi_{\varepsilon}(u^*))dt \le C,$$

by (2.7) and (2.11). The uniform coercivity of J_{ε} with respect to ε, implies that $\{u_{\varepsilon}\}$ is bounded in $L^2(0,T;U)$ and let $u_{\varepsilon} \to u^0$ on a subsequence, weakly in $L^2(0,T;U)$. By Lemma 2.1 $y_{\varepsilon} \to y^0 = \theta(u^0)$ strongly in $C(0,T;H)$. Passing to the limit in the above inequality, we get

$$\int_0^T (1/2|y^0 - y_d|_H^2 + 1/2|u^0 - u^*|_U^2 + \psi(u^0))dt \le \int_0^T (1/2|y^* - y_d|_H^2 + \psi(u^*))dt.$$

Since $[y^*,u^*]$ is an optimal pair, it yields that $u^0 = u^*$, $y^0 = y^*$ and (2.13), (2.14).

Lemma 2.3. The mapping θ_ε is Gateaux differentiable on $L^2(0,T;H)$ and $r = \nabla\theta_\varepsilon(g)w$ satisfies

(2.15) $\quad \dot\beta^\varepsilon(\theta_\varepsilon(g))r - \Delta\int_0^t r\,ds = \int_0^t w\,ds \qquad$ in Q,

(2.16) $\quad |\nabla\theta_\varepsilon(g)w|_{L^2(0,t;H)} \le C\int_0^t |w(s)|_{V^*}ds, \qquad t\in[0,T]$,

for $g, w \in L^2(0,T;H)$.

Proof.

We put $y^\lambda = \theta_\varepsilon(g+\lambda w)$, $v^\lambda = \beta^\varepsilon(y^\lambda)$, $y = \theta_\varepsilon(g)$, $v = \beta^\varepsilon(y)$ and we have

(2.17) $\quad v^\lambda - v - \Delta\int_0^t(y^\lambda - y) = \lambda\int_0^t w.$

We multiply by $y^\lambda - y$ and we use (2.6)

(2.18) $\quad |y^\lambda(t) - y(t)|_H^2 - 1/2\, d/dt(\Delta\int_0^t(y^\lambda - y), \int_0^t(y^\lambda - y))_H = \lambda(\int_0^t w, y^\lambda - y)_H.$

Let $z^\lambda = (y^\lambda - y)/\lambda$. Then, by (2.18) we get $\{\int_0^t z^\lambda\}$ bounded in $L^\infty(0,T;V)$, $\{z^\lambda\}$ bounded in $L^2(0,T;H)$, so

$\int_0^t z^\lambda \to z \qquad$ strongly in $C(0,T;H)$,

$\qquad\qquad\qquad$ weakly* in $L^\infty(0,T;V)$,

$z^\lambda \to z_t \qquad$ weakly in $L^2(0,T;H)$.

Since β^ε is Lipschitzian (ε is fixed) we see that $\{(v^\lambda - v)/\lambda\}$ is bounded in $L^2(0,T;H)$. We remark that $y^\lambda \to y$ strongly in $L^2(0,T;H)$ and the Lebesgue theorem shows that

$$(v^\lambda - v)/\lambda \to \dot\beta^\varepsilon(y)z_t$$

weakly in $L^2(0,T;H)$. We divide by λ in (2.17) and we denote $r = z_t$. Passing to the limit, we obtain (2.15). As (2.15), with zero Dirichlet conditions for $z = \int_0^t r \in L^\infty(0,T;V)$, has a unique solution, we see that z^λ is convergent without taking subsequences. As concerns (2.16), we start from

$$v_t^\lambda - v_t - \Delta(y^\lambda - y) = \lambda w$$

and we multiply by $v^\lambda - v$ in the scalar product of V^*. Integrating over $[0,t]$, we infer

$$1/2|v^\lambda - v|_{V^*}^2 + \int_0^t |y^\lambda - y|_H^2 \le \lambda C(\int_0^t w, v^\lambda - v)_{V^*}.$$

Remark 2.4. By virtue of (2.16), the operator $\nabla\theta_\varepsilon(g)$ may be extended by continuity to the whole $L^1(0,T;V^*)$, with values in $L^2(0,T;H)$, preserving the same inequality.

We may define the adjoint operator $\nabla\theta_\varepsilon(g)^*: L^2(0,T,H)\to L^\infty(0,T;V)$ and for any $q,g\in L^2(0,T;H)$ we have the inequality

(2.19) $$|\nabla\theta_\varepsilon(g)^* q|_{L^\infty(t,T;V)} \leq C|q|_{L^2(t,T;H)}.$$

The proof is based on the definition of the adjoint and it is similar with (1.14), Ch.II.

Lemma 2.5. For any $\varepsilon > 0$ there is $p_\varepsilon \in L^\infty(0,T;V)$ such that

(2.20) $$p_\varepsilon = -\nabla\theta_\varepsilon(Bu_\varepsilon + f)^*(y_\varepsilon - y_d),$$

(2.21) $$B^* p_\varepsilon = \partial\Psi_\varepsilon(u_\varepsilon) + u_\varepsilon - u^*.$$

The argument is the same as in Lemma 1.3, II, and we leave it to the reader.

Remark 2.6. The above results correspond to the conditions (a), (b), (c) from Chapter II, § 1. Now we can prove the proposition

Proposition 2.7. There is $p_\varepsilon \in L^\infty(0,T;V)$ such that:

(2.22) $$\dot{\beta}^\varepsilon(y_\varepsilon)(p_\varepsilon)_t + \Delta p_\varepsilon = y_\varepsilon - y_d \quad \text{in } Q,$$

(2.23) $$p_\varepsilon(T,x) = 0 \quad \text{in } \Omega,$$

and (2.21) are satisfied. Moreover, $y_\varepsilon \to y^*$ strongly in $C(0,T;H)$, $u_\varepsilon \to u^*$ strongly in $L^2(0,T;U)$, $p_\varepsilon \to p^*$ strongly in $C(0,T;H)$, weakly* in $L^\infty(0,T;V)$, $(p_\varepsilon)_t \to p_t^*$ weakly in $L^2(0,T;H)$ and

(2.24) $$B^* p^*(t) \in \partial\Psi(u^*(t)) \quad \text{in } [0,T].$$

Proof

The equation (2.22), (2.23) has a unique strong solution by standard results for linear nondegenerate parabolic equations. The equivalence with (2.20) is obtained by the definition of the adjoint.

We multiply (2.22) by $(p_\varepsilon)_t$ and we remark that $\dot{\beta}^\varepsilon(y_\varepsilon) \geq 1$ in Q. We deduce that $\{p_\varepsilon\}$ is bounded in $L^\infty(0,T;V)$, $\{(p_\varepsilon)_t\}$ is bounded in $L^2(0,T;H)$ and the convergences $p_\varepsilon \to p^*$ strongly in $C(0,T;H)$, $(p_\varepsilon)_t \to p_t^*$ weakly in $L^2(0,T;H)$. The relation (2.24) is a consequence of the demiclosedness of $\partial\Psi$.

Remark 2.8. In some important cases one may obtain the optimality system satisfied by u^*, y^*, p^*.

First, let us assume that β satisfies

(2.25) $$\beta \text{ is locally Lipschitzian and } |\dot{\beta}(y)| \leq C(|\beta(y)| + |y| + 1) \text{ a.e.} R,$$

(2.26) $$\beta = \xi - \mu,$$

where $\xi,\mu : R\to R$ are convex mappings defined on the whole real axis.

We remark that these hypotheses are of the same type as in Chapter II.

Proposition 2.9. Under the above conditions, if $[y^*,u^*]$ is an optimal pair for (2.1)-(2.5), there are $p^* \in W^{1,2}(0,T;H)\cap L^\infty(0,T;V)$, $h\in L^1(Q)$ such that

(2.27) $h + \Delta p^* = y^* - y_d$ in Q,

(2.28) $\dot\beta^\varepsilon(y_\varepsilon)(p_\varepsilon)_t \to h$ weakly in $L^1(Q)$.

Proof

Multiplying (2.22) by $(p_\varepsilon)_t$ we also get that $\{\dot\beta^\varepsilon(y_\varepsilon)(p_\varepsilon)_t^2\}$ is bounded in $L^1(Q)$. By (2.25) and the boundedness of $\{\beta^\varepsilon(y_\varepsilon)\}$ in $L^2(Q)$, it yields that $\{\dot\beta^\varepsilon(y_\varepsilon)\}$ is bounded in $L^2(Q)$. Then, the Young's inequality shows that $\{\dot\beta^\varepsilon(y_\varepsilon)(p_\varepsilon)_t\}$ is bounded in $L^s(Q)$, for some $s > 1$, as in Ch.II, § 3.

Denoting by h the weak limit in $L^s(Q)$ of the sequence $\{\dot\beta^\varepsilon(y_\varepsilon)(p_\varepsilon)_t\}$, we obtain (2.27), (2.28).

Theorem 2.10. Under hypotheses (2.25), (2.26), there is $p^* \in L^\infty(0,T;V) \cap W^{1,2}(0,T;H)$ which satisfies together with u^*, y^*, the equation

$$D\beta(y^*)p_t^* + \Delta p^* = y^* - y_d \quad \text{in } Q,$$
$$p^*(T,x) = 0 \quad \text{in } \Omega,$$
$$p^*(t,x) = 0 \quad \text{in } \Sigma,$$

in a generalized sense.

Here $D\beta$ is the Clarke generalized gradient of the locally Lipschitzian mapping β.

The proof is based on the condition (2.26) and on the use of a saddle function as in § 3, Chapter II. For more details, we quote the paper [114].

Now, we consider again the two-phase Stefan problem, that is we assume that β is of the form (1.4). Obviously, β doesn't satisfy (2.25), but we remark that it is sublinear

(2.29) $|\beta(y)| \le C(|y| + 1), \ y \in R.$

We impose the supplementary condition

(2.30) $\text{mes}\{(t,x) \in Q;\ y^*(t,x) = 0\} = 0,$

that is, we suppose that the free boundary has zero measure.

Theorem 2.11. Under assumptions (2.30), (1.4) there is $p^* \in L^\infty(0,T;V) \cap W^{1,2}(0,T;H)$ satisfying the optimality system

(2.31) $\dot\beta(y^*)p_t^* + \Delta p^* = y^* - y_d$ in Q,

(2.32) $p^*(T,x) = 0$ in Ω,

(2.33) $p^*(t,x) = 0$ on Σ.

We remark that (2.31) has sense under condition (2.30).

Proof.

We have to pass to the limit in (2.22). A detailed calculation gives

$$(2.34)\qquad \dot{\gamma}_\varepsilon(r) = \begin{cases} 0 & r < 0, \\ 1/\varepsilon & 0 < r < \varepsilon L, \\ m/(1+\varepsilon m) & r > \varepsilon L, \end{cases}$$

where we take, for the sake of simplicity, $c_1 = 1$, $c_2 > 1$, $c_2 - 1 = m$. We deduce that

$$(2.35)\qquad r\dot{\gamma}_\varepsilon(r) = \gamma_\varepsilon(r) - g_\varepsilon(r)$$

$$(2.36)\qquad g_\varepsilon(r) = \begin{cases} 0 & r < \varepsilon L, \\ L/(1+\varepsilon m) & r > \varepsilon L. \end{cases}$$

It yields that

$$y\dot{\beta}^\varepsilon(y) = y(1 + \dot{\gamma}^\varepsilon(y)) = y + y\int_{-1}^{1} \dot{\gamma}_\varepsilon(y - \varepsilon^2 s)\rho(s)ds = y + \varepsilon^2\int_{-1}^{1} \dot{\gamma}_\varepsilon(y - \varepsilon^2 s)\rho(s)ds +$$

$$+ \int_{-1}^{1} \dot{\gamma}_\varepsilon(y - \varepsilon^2 s)(y - \varepsilon^2 s)\rho(s)ds = y + h_\varepsilon(y) - g^\varepsilon(y) + \gamma^\varepsilon(y) = \beta^\varepsilon(y) + h_\varepsilon(y) - g^\varepsilon(y),$$

where

$$(2.37)\qquad g^\varepsilon(y) = \int_{-1}^{1} g_\varepsilon(y - \varepsilon^2 s)\rho(s)ds,$$

$$(2.38)\qquad h_\varepsilon(y) \to 0 \qquad \text{uniformly with respect to } y,$$

since γ_ε is Lipschitzian of constant $1/\varepsilon$.

Obviously $y_\varepsilon \to y^*$ a.e. Q. Then, (2.30) implies $\beta^\varepsilon(y_\varepsilon) \to \beta(y^*)$ a.e. Q and, by (2.29), we get that $\beta^\varepsilon(y_\varepsilon) \to \beta(y^*)$ strongly in $L^2(Q)$.

Moreover, the above calculation gives

$$(2.39)\qquad y_\varepsilon \dot{\beta}^\varepsilon(y_\varepsilon)(p_\varepsilon)_t = \beta^\varepsilon(y_\varepsilon)(p_\varepsilon)_t + h_\varepsilon(y_\varepsilon)(p_\varepsilon)_t - g^\varepsilon(y_\varepsilon)(p_\varepsilon)_t$$

$$(2.40)\qquad y_\varepsilon \dot{\beta}^\varepsilon(y_\varepsilon)(p_\varepsilon)_t \to \beta(y^*)p_t^* - w$$

weakly in $L^1(Q)$, where w is the weak limit in $L^2(Q)$ (on a subsequence) of $g^\varepsilon(y_\varepsilon)(p_\varepsilon)_t$.

We subtract two approximating state equations corresponding to the parameters $\varepsilon, \lambda > 0$ and we multiply by $y_\varepsilon - y_\lambda$. Integrating by parts, we obtain

$$\int_\Omega (\beta^\varepsilon(y_\varepsilon(T,x)) - \beta^\lambda(y_\lambda(T,x)))(y_\varepsilon(T,x) - y_\lambda(T,x))dx - \int_0^T\int_\Omega (\beta^\varepsilon(y_\varepsilon) - \beta^\lambda(y_\lambda))((y_\varepsilon)_t -$$

$$- (y_\lambda)_t)dxdt + \int_0^T\int_\Omega |\operatorname{grad}(y_\varepsilon - y_\lambda)|^2 dxdt = \int_0^T B(u_\varepsilon - u_\lambda)(y_\varepsilon - y_\lambda)dxdt.$$

By <u>Lemma 2.1</u> and the strong convergence of $\beta^\varepsilon(y_\varepsilon)$ in $L^2(0,T;H)$ we obtain that $y_\varepsilon \to y^*$ strongly in $L^2(0,T;V)$. The adjoint equation (2.22) shows that $\dot{\beta}^\varepsilon(y_\varepsilon)(p_\varepsilon)_t \to 1$ weakly* in $L^\infty(0,T;V^*)$. It yields that

(2.41) $y_\varepsilon \dot{\beta}^\varepsilon(y_\varepsilon)(p_\varepsilon)_t \to y^* \cdot l$

at least in $\mathcal{D}'(Q)$. In fact, according to (2.40), the convergence is true in the weak topology of $L^1(Q)$. We use (2.36) and we deduce that $g^\varepsilon(y_\varepsilon) \to g(y^*)$ strongly in $L^2(Q)$, where:

$$g(y) = \begin{cases} 0 & y < 0 \\ L & y > 0. \end{cases}$$

Then $w(t,x) = g(y^*(t,x))p_t^*(t,x)$ a.e.Q and we have the identity

$$\beta(y^*)p_t^* - g(y^*)p_t^* = y^* l \qquad \text{a.e.} Q.$$

Under condition (2.30), a simple calculation, allows to rewrite this identity as follows

$$y^* \dot{\beta}(y^*)p_t^* = y^* l \qquad \text{a.e.} Q$$

and we obtain the desired conclusion

$$l = \dot{\beta}(y^*)p_t^* \qquad \text{a.e.} Q.$$

Remark 2.12. If $U = H$, $B = I$ and no control constraints are imposed, then (2.24) implies $u^* \in L^\infty(0,T;H^1(\Omega))$, for quadratic cost functionals. This regularity result is preserved for certain control constraints, by Remark 2.6, Ch.II.

3. Boundary control

In this section we study the unconstrained control problem

(3.1) Minimize $\int_0^T (1/2|y - y_d|_H^2 + 1/2|u|_{L^2(\Gamma)}^2)dt$,

(3.2) $v_t - \Delta y = 0$ in Q,

(3.3) $v \in \beta(y)$ in Q,

(3.4) $\partial y/\partial n = u$ in Σ,

(3.5) $v(0,x) = v_0(x)$ in Ω.

Here $\partial y/\partial n$ is the normal derivative of y to Γ, $u \in L^2(\Gamma)$ is the control, and β, y_d, v_0 are as in the previous section.

A similar problem was considered by C.Saguez [109], Ch.4, by a semidiscretization method. In the works of I.Pawlow [91], [93], M.Niezgodka and I.Pawlow [84], T.Roubicek [102] the case of differentiable, with respect to t, boundary control is analysed, for more complex models of the melting process.

It is to be remarked that in this section and in the next one we use a standard penalization approach, not the adapted penalization method. This has the advantage to allow numerical studies as well, which is not possible for the adapted variant which involves the solution u^* of the original control problem, in the definition of the approximating problem. However, on the theoretical level, the adapted penalization method provides better convergence results, which are useful in the study of the optimality conditions, for instance.

Let $\theta : L^2(\Sigma)\to L^2(Q)$ be the mapping $u\to y$ defined by (3.2) - (3.5). The proof of the existence of optimal pairs in $L^2(\Sigma)$ for the problem (3.1) - (3.5) is based on a nonstandard result on the continuous dependence in the weak topology for (3.2) - (3.5). We use the notations of Proposition 1.4.

Proposition 3.1. Let $u_n\to u$ weakly in $L^2(\ )$. Then $y_n = \theta(u_n)\to y = \theta(u)$ weakly in $L^2(0,T;H^1(\Omega))$.

Proof.

We have

(3.6) $$\tilde\beta(dw_n/dt) + \tilde A w_n \ni f_n$$
$$w_n(0) = 0$$

and $w_n = \int_0^t y_n ds$, $f_n = \int_0^t g_n ds + v_0$, with g_n defined by

(3.7) $$\int_0^T (g_n,\Psi)_{H^1(\Omega)\times H^1(\Omega)^*} = \int_0^T\int_\Gamma u_n\Psi,\quad \Psi\in L^2(0,T;H^1(\Omega)).$$

Obviously $f_n\to f$ (defined in a similar way, but starting from u) weakly in $W^{1,2}(0,T;H^1(\Omega)^*)$.

As in Proposition 1.4, we see that $\{w_n\}$ is bounded in $L^\infty(0,T;H^1(\Omega))$, $\{dw_n/dt\}$ is bounded in $L^2(0,T;H)$ and $\{\tilde\beta(dw_n/dt)\}$ is bounded in $L^2(0,T;H)$.

Let $\bar A : H^1(\Omega)\to H^1(\Omega)^*$ be given by $\bar A v = v + \tilde A v$ and let $v_n\in L^2(0,T;H^1(\Omega))$ satisfy

(3.8) $$\bar A v_n = f_n - \tilde A w_n.$$

The existence of v_n is a consequence of the monotonicity and coercivity of $\bar A$.

Then $v_n = h_n - w_n$, where $h_n = \bar A^{-1}(f_n + w_n)$ and $h_n\to h$ weakly in $W^{1,2}(0,T;H^1(\Omega))$.

The relation (3.6) is equivalent with

(3.9) $$dv_n/dt + \tilde\beta^{-1}\bar A v_n = dh_n/dt,$$
$$v_n(0) = \bar A^{-1}(v_0) = \tilde v_0.$$

According to Barbu [12], p.214, the operator $\tilde\beta^{-1}\bar A$ is a subdifferential in $H^1(\Omega)\times H^1(\Omega)$. We remark that v_0 is in the domain of $\tilde\beta^{-1}A$ and standard results for abstract evolution problems show that (3.9) has a unique solution $v_n\in C(0,T;H^1(\Omega))$, $dv_n/dt\in L^2(0,T;H^1(\Omega))$. We multiply (3.9) by $\bar A\, dv_n/dt$

$$(3.10)\qquad \int_0^T (dv_n/dt, \bar{A}\, dv_n/dt)_{H^1(\Omega)\times H^1(\Omega)^*}dt + \int_0^T (\partial\varphi^*(\bar{A}v_n), d/dt\bar{A}v_n)_{H^1(\Omega)\times H^1(\Omega)^*}dt = \int_0^T (dh_n/dt, \bar{A}\, dv_n/dt)dt.$$

Here $\varphi: H^1(\Omega)\to]-\infty,+\infty]$ is such that $\tilde{\beta}|_{H^1(\Omega)} = \partial\varphi$, so $\tilde{\beta}^{-1} = \partial\varphi^*$, where φ^* is the convex conjugate of φ. We obtain that $\{dv_n/dt\}$ is bounded in $L^2(0,T;H^1(\Omega))$ and, consequently, $\{w_n\}$ is bounded in $W^{1,2}(0,T;H^1(\Omega))$.

Obviously, by (3.6), we have that $\{f_n - Aw_n\}$ is bounded in $L^2(0,T;H)$ and

$$d/dt(f_n - \tilde{A}w_n) = df_n/dt - \tilde{A}dw_n/dt$$

is bounded in $L^2(0,T;H^1(\Omega)^*)$. The Aubin compactness result shows that $f_n - \tilde{A}w_n \to f - \tilde{A}w$ strongly in $L^2(0,T;H^1(\Omega)^*)$, where $w_n \to w$ weakly* in $L^\infty(0,T;H^1(\Omega))$ and $dw_n/dt \to dw/dt$ weakly in $L^2(0,T;H^1(\Omega))$. It yields that $\tilde{\beta}(dw_n/dt)\to d$ strongly in $L^2(0,T;H^1(\Omega)^*)$ and, by the demiclosedness of $\tilde{\beta}$, we have $d\in\tilde{\beta}(dw/dt)$. Finally, we may pass to the limit in (3.6) and infer that $y = dw/dt = \theta(u)$, which ends the proof.

Theorem 3.2. The problem (3.1) - (3.5) has at least one optimal pair $[y^*,u^*]$ in $L^2(0,T;H^1(\Omega)) \times L^2(\Sigma)$.

Proof.

The cost functional (3.1) is weakly lower semicontinuous and coercive on $L^2(\Sigma)$.

Remark 3.3. The functional (3.1) isn't convex with respect to u, if we replace $y = \theta(u)$.

We define the approximate problem

$$(3.11)\qquad \text{Minimize} \int_0^T (1/2\,|y-y_d|_H^2 + 1/2\,|u|^2_{L^2(\Gamma)})dt,$$

$$(3.12)\qquad \partial\beta^\varepsilon(y)/\partial t - \Delta y = 0 \quad \text{in } Q,$$

$$(3.13)\qquad \partial y/\partial n = u \quad \text{in } \Sigma,$$

$$(3.14)\qquad y(0,x) = y_0(x) \quad \text{in } \Omega,$$

where β^ε is given by (2.6) and $y_0 = \tilde{\beta}^{-1}(v_0)$.

The solution of (3.12) - (3.14) may be understood in the variational sense too and, obviously, the control problem (3.11)- (3.14) has at least one optimal pair $[y_\varepsilon, u_\varepsilon]$. Let $\theta_\varepsilon: L^2(\Sigma)\to L^2(Q)$ be the mapping $u\to y$ given by (3.12) - (3.14).

Lemma 3.4. For any $u\in L^2(\Sigma)$ there exists the linear operator $\nabla\theta_\varepsilon(u): L^2(\Sigma)\to L^2(Q)$ defined by

$$(3.15)\qquad \nabla\theta_\varepsilon(u)v = \lim_{\lambda\to 0}(\theta_\varepsilon(u+\lambda v) - \theta_\varepsilon(u))/\lambda$$

in the weak topology of $L^2(Q)$, for any $v\in L^2(\Sigma)$. Moreover $\nabla\theta_\varepsilon(u)v = dz/dt$, where z satisfies

$$(3.16)\qquad \dot{\beta}^\varepsilon(\theta_\varepsilon(u))dz/dt + \tilde{A}z = h \text{ in } Q,$$

$$(3.17)\qquad z(0) = 0,$$

and h is explained below.

Proof.

Let $y_\lambda = \theta_\varepsilon(u+\lambda v)$, $y = \theta_\varepsilon(u)$ and $h = \int_0^t K\,ds$ with K obtained by (3.7), starting from $v \in L^2(\Sigma)$. We subtract the equations corresponding to y_λ, y and we multiply by $dw_\lambda/dt - dw/dt$, where $w_\lambda = \int_0^t y_\lambda\,ds$, $w = \int_0^t y\,ds$:

$$\int_0^t |dw_\lambda/dt - dw/dt|_H^2 ds + 1/2(\tilde{A}(w_\lambda(t) - w(t)), w_\lambda(t) - w(t))_{H^1(\Omega)\times H^1(\Omega)^*} \le$$

$$\le \lambda \int_0^t (h, dw_\lambda/dt - dw/dt)ds.$$

Integrating by parts in the right-hand side and denoting $z_\lambda = (w_\lambda - w)/\lambda$, we get $\{z_\lambda\}$, $\{dz_\lambda/dt\}$ bounded in $L^\infty(0,T;H^1(\Omega))$, $L^2(0,T;H)$ respectively. Furthermore, $w_\lambda \to w$ strongly in $L^\infty(0,T;H^1(\Omega))$, $dw_\lambda/dt \to dw/dt$ strongly in $L^2(0,T;H)$. Since β^ε is Lipschitzian, one may show in a usual way that

$$(\beta^\varepsilon(dw_\lambda/dt) - \beta^\varepsilon(dw/dt))/\lambda \to \dot{\beta}^\varepsilon(dw/dt)dz/dt$$

weakly in $L^2(0,T;H)$ and (3.16), (3.17) follow easily.

Lemma 3.5. For any $\varepsilon > 0$ there is $p_\varepsilon \in L^\infty(0,T;H^1(\Omega)) \cap W^{1,2}(0,T;H)$ such that

(3.18) $\dot{\beta}^\varepsilon(y_\varepsilon)\partial p_\varepsilon/\partial t - \tilde{A}p_\varepsilon = y_\varepsilon - y_d$ in Q,

(3.19) $p_\varepsilon(T,x) = 0$ in Ω,

(3.20) $p_\varepsilon(t,x) = u_\varepsilon(t,x)$ in Σ.

Proof

The adjoint approximating system (3.18), (3.19) is equivalent with the nondegenerate parabolic problem

$$\dot{\beta}^\varepsilon(y_\varepsilon)\partial p_\varepsilon/\partial t + \Delta p_\varepsilon = y_\varepsilon - y_d \quad \text{in } Q,$$
$$p_\varepsilon(T,x) = 0 \quad \text{in } \Omega,$$
$$\partial p_\varepsilon/\partial n = 0 \quad \text{in } \Sigma.$$

The existence of a unique generalized solution $p_\varepsilon \in L^\infty(0,T;H^1(\Omega)) \cap W^{1,2}(0,T;L^2(\Omega))$ follows easily (V.Mikhailov [77], Ch.VI).

By the definition of the adjoint, it yields that we have

$$\nabla\theta_\varepsilon(u_\varepsilon)^*(y_\varepsilon - y_d) = -p_\varepsilon|_\Sigma.$$

The Gateaux differential of the cost functional (3.11) vanishes in the minimum point $u_\varepsilon \in L^2(\Sigma)$ and, by the above remark, we get (3.20). See Tiba [115] for a more detailed argument.

Lemma 3.6. We suppose that $u_\varepsilon \to u$ weakly in $L^2(\Sigma)$; then $\theta_\varepsilon(u_\varepsilon) \to \theta(u)$ weakly in

$L^2(0,T;H^1(\Omega))$.

This is a variant of Proposition 3.1.

Remark 3.7. Consequently, $\theta_\varepsilon(u) \rightharpoonup \theta(u)$ weakly in $L^2(0,T;H^1(\Omega))$, for every $u \in L^2(\Sigma)$. Using the strong monotonicity of β, one may show that $\theta_\varepsilon(u) \to \theta(u)$ strongly in $L^2(Q)$.

Proposition 3.8. On a subsequence, we have

(3.21) $\quad u_\varepsilon \to u^*$ strongly in $L^2(\Sigma)$,

(3.22) $\quad y_\varepsilon \to y^*$ strongly in $L^2(Q)$,

(3.23) $\quad p_\varepsilon \to p^*$ strongly in $C(0,T;H)$.

Proof.

Since we don't use the adapted penalization method here, the argument is different. We denote J_ε the functional (3.11). For any $u \in L^2(\Sigma)$, we have the inequality

(3.24) $$J_\varepsilon(y_\varepsilon, u_\varepsilon) \le J_\varepsilon(\theta_\varepsilon(u), u) = \int_0^T (1/2|\theta_\varepsilon(u) - y_d|_H^2 + 1/2|u|^2_{L^2(\Gamma)}).$$

According to Remark 3.7 we deduce that $\{u_\varepsilon\}$ is bounded in $L^2(\Sigma)$ and, on a subsequence, $u_\varepsilon \to \bar{u}$ weakly in $L^2(\Sigma)$, Lemma 3.6. implies $\theta_\varepsilon(u_\varepsilon) \rightharpoonup \theta(\bar{u})$ weakly in $L^2(0,T;H^1(\Omega))$ and we may pass to the limit in (3.24):

$$\int_0^T (1/2|\theta(\bar{u}) - y_d|_H^2 + 1/2|\bar{u}|^2_{L^2(\Gamma)})dt \le \int_0^T (1/2|\theta(u) - y_d|_H^2 + 1/2|u|^2_{L^2(\Gamma)})dt$$

so $[\bar{y},\bar{u}]$, $\bar{y} = \theta(\bar{u})$, is an optimal pair for the problem (3.1) - (3.5), which we denote $[y^*,u^*]$.

Here, we also use the fact that $\theta_\varepsilon(u) \to \theta(u)$ strongly in $L^2(0,T;H)$, for any $u \in L^2(\Sigma)$. This may be obtained as follows. By (2.6) and (3.12), (1.32), a simple calculus gives:

$$\int_0^T |\theta_\varepsilon(u) - \theta_\lambda(u)|_H^2 dt + \int_0^T (\gamma^\varepsilon(\theta_\varepsilon(u)) - \gamma^\lambda(\theta_\lambda(u)), \theta_\varepsilon(u) - \theta_\lambda(u))_H dt +$$
$$+ \int_0^T (A\int_0^t [\theta_\varepsilon(u) - \theta_\lambda(u)]d\tau, \theta_\varepsilon(u) - \theta_\lambda(u))_H dt = 0.$$

Since

$$(\gamma_\varepsilon(\theta_\varepsilon(u)) - \gamma_\lambda(\theta_\lambda(u)), \theta_\varepsilon(u) - \theta_\lambda(u))_H \ge (\gamma_\varepsilon(\theta_\varepsilon(u)) - \gamma_\lambda(\theta_\lambda(u)),$$
$$\varepsilon\gamma_\varepsilon(\theta_\varepsilon(u)) - \lambda\gamma_\lambda(\theta_\lambda(u)))_H$$

the boundedness of $\{\theta_\varepsilon(u)\}$ and $\{\beta^\varepsilon(\theta_\varepsilon(u))\}$ in $L^2(0,T;H)$, show that $\{\theta_\varepsilon(u)\}$ is a Cauchy sequence in $L^2(0,T;H)$.

We multiply (3.18) by dp_ε/dt and we integrate over $[t,T]$. It yields that $\{p_\varepsilon\}$ is bounded in $L^\infty(0,T;H^1(\Omega))$ and $\{dp_\varepsilon/dt\}$ is bounded in $L^2(Q)$. The Aubin theorem [2] gives $p_\varepsilon \to p^*$ strongly in $L^2(0,T;H^{3/4}(\Omega))$ and, by the trace theorem, we get $u_\varepsilon \to u^*$ strongly in $L^2(\Sigma)$, according to (3.20).

The relations (3.22), (3.23) are obvious, now.

Corollary 3.9. Let us denote by J the functional (3.1). Then:

(3.25) $$\lim_{\varepsilon\to 0} J(\theta(u_\varepsilon),u_\varepsilon) = J(y^*,u^*).$$

Proof.

By (3.21), (3.22) we have $\lim_{\varepsilon\to 0} J_\varepsilon(y_\varepsilon,u_\varepsilon) = J(y^*,u^*)$. By the form of J_ε, J, it is enough to show that $\lim_{\varepsilon\to 0}\theta(u_\varepsilon) = \theta(u^*) = y^*$ strongly in $L^2(Q)$. We know that $\theta(u_\varepsilon)\to y^* = \theta(u^*)$ weakly in $L^2(0,T;H^1(\Omega))$ by Proposition 3.1.

We subtract the equations corresponding to u_ε, u^* and we multiply by $y^\varepsilon - y^*$ ($y^\varepsilon = \theta(u_\varepsilon)$):

$$|y^\varepsilon - y^*|^2_{L^2(Q)} + (\tilde{A}w^\varepsilon - \tilde{A}w^*, y^\varepsilon - y^*)_{L^2(Q)} = \int_0^T (f_\varepsilon - f^*, y^\varepsilon - y^*)_{H^1(\Omega)\times H^1(\Omega)^*},$$

where $f_\varepsilon = \int_0^t g_\varepsilon$, $f^* = \int_0^t g^*$, $w^\varepsilon = \int_0^t y^\varepsilon$, $w^* = \int_0^t y^*$ and g_ε, g^* are obtained by (3.7), starting from u_ε, u^*.

Obviously $f_\varepsilon\to f^*$ strongly in $W^{1,2}(0,T;H^1(\Omega)^*)$ and, from the boundedness of $\{y^\varepsilon\}$ in $L^2(0,T;H^1(\Omega))$, we get the desired conclusion.

Remark 3.10. It yields that u_ε is a suboptimal control for the problem (3.1) - (3.5).

We close this section with a partial answer to the question of the first order optimality conditions for the problem (3.1) - (3.5). We impose again the assumption (2.30).

Proposition 3.11. Under hypothesis (2.30) we have

(3.26) $$y_\varepsilon\dot{\beta}^\varepsilon(y_\varepsilon)\partial p_\varepsilon/\partial t\to y^*\dot{\beta}(y^*)\partial p^*/\partial t$$

weakly in $L^1(Q)$.

Proof

A calculation similar to Thm.2.11 gives

$$y_\varepsilon\dot{\beta}^\varepsilon(y_\varepsilon)\partial p_\varepsilon/\partial t = \beta^\varepsilon(y_\varepsilon)\partial p_\varepsilon/\partial t + h_\varepsilon(y_\varepsilon)\partial p_\varepsilon/\partial t - g^\varepsilon(y_\varepsilon)\partial p_\varepsilon/\partial t,$$

in the notations of (2.39).

Since $y_\varepsilon\to y^*$ a.e.Q, (2.30) and (1.4) imply that $\beta^\varepsilon(y_\varepsilon)\to\beta(y^*)$ a.e.Q and, by (2.29), we have $\beta^\varepsilon(y_\varepsilon)\to\beta(y^*)$ strongly in $L^2(Q)$. It yields that the right-hand side is weakly convergent in $L^1(Q)$ to $\beta(y^*)\partial p^*/\partial t - g(y^*)\partial p^*/\partial t$, which gives (3.26) by a direct reasonning.

Remark 3.12. We know that $y_\varepsilon\to y^*$ weakly in $L^2(0,T;H^1(\Omega))$ and (3.18) implies that $\dot{\beta}^\varepsilon(y_\varepsilon)\partial p_\varepsilon/\partial t\to l$ weakly in $L^2(0,T;H^1(\Omega)^*)$. Combining these facts with Proposition 3.11, we make the conjecture that $l = \dot{\beta}(y^*)\partial p^*/\partial t$, i.e. the adjoint optimality system is

$$\dot{\beta}(y^*)dp^*/dt - \tilde{A}p^* = y^* - y_d \text{ in } Q,$$

$$p^*(T) = 0.$$

Remark 3.13. Obviously $p^* \in L^\infty(0,T;H^1(\Omega))$ and, by (3.20), passing to the limit, we obtain that $u^* \in L^\infty(0,T;H^{1/2}(\Gamma))$, a regularity result for the optimal controls of the problem (3.1) - (3.5).

4. Discretization

In this section we give a detailed justification of a computational procedure used to find a suboptimal control in the problem (3.1) - (3.5).

Similar results may be obtained in the completion of Chapter III, section 4 and we quote the work of Neittaanmaki and Tiba [130] in this respect. The literature on the approximation of control problems governed by variational inequalities or free boundary problems is rich: V.Arnautu [10], [9], J.P.Yvon [144], J.F.Bonnans [22], C.Saguez [109], K.-H.Hoffmann and J.Sprekels [60], J.Hlavacek, J.Bock, J.Lovisek [58]. See also the recent books of I. Pawlow [90] and Haslinger and Neittaanmaki [59] and their references.

We begin by discussing the discretization of two-phase Stefan problems by the finite element method in space and by finite differences in time. An important role is played by the notion of weak solution (see section 1) and, due to the strong relationship between the weak and the variational solution, the results which follow will be useful in the discretization of the associated the control problem too.

We fix our attention on the Neumann problem. In the case when the boundary data are differentiable with respect to the time, error estimates were obtained by Pawlow [94], Jerome and Rose [61]. General boundary conditions were considered by White [138], [139] by a finite differences method.

We analyse first the influence of the regularity of the boundary data on the convergence properties of the sequence of discrete solutions. This is important from the point of view of the applications to control problems, since, generally, one works with integrable controls only.

For the sake of simplicity, let Ω be a polygonal domain in R^2. The results are also valid for $\Omega \subset R^N$. The Ω is covered by a union of triangles $\mathcal{T}^h$, where h is the length of the largest edge and $\mathcal{T}^h$ satisfies the conditions of Zlamal [143]. See [34], [52], [87] for an introduction in the theory and in the numerical applications of the finite element method.

Let V_h be the space of continuous functions, linear on each triangle of $\mathcal{T}^h$, with the norm $|\cdot|_h$ induced by the modified $L^2(\Omega)$ inner product

$$(4.1) \qquad (w,v)_h = 1/3 \sum_{i \leq I} W_i v_i w_i, \quad v,w \in V_h.$$

Here I is the number of vertices associated with $\mathcal{T}^h$, W_i is the sum of the areas of the triangles with a vertex in i and v_i, w_i are the values of v,w at the node i.

V_h is a finite dimensional space of dimension I and there is C > 0, independent of h, such that

$$(4.2) \qquad |v|_H \leq |v|_h \leq C|v|_H, \quad v \in V_h,$$

(4.3) $$|(w,v)_h - (w,v)_H| \leq Ch^2 |w|_V \cdot |v|_V \leq Ch|v|_H \cdot |w|_V, \quad w,v \in V.$$

We denote $H = L^2(\Omega)$, $V = H^1(\Omega)$.

Let L_h be the space of restrictions to $\partial\Omega$ of the functions from V_h, endowed with the norm of $L^2(\partial\Omega)$. L_h has the dimension $J < I$, the number of nodes on $\partial\Omega$.

We assume that the interval $[0,T]$ is divided in m equal subintervals of length $K > 0$, $m.K = T$. The approximating equations are

(4.4) $$((v^{n+1} - v^n)/k, v)_h + \int_\Omega \text{grad } y^{n+1} . \text{grad } v - \int_{\partial\Omega} u^{n+1} v = (f^{n+1}, v)_h,$$

(4.5) $$v_i^n \in \beta(y_i^n), \quad i \leq I,$$

for all $v \in V_h$, $0 \leq n \leq m-1$.

To avoid a tedious argumentation, we suppose that the data are defined pointwise and f_i^n, u_i^n denote the value of f, u at the moment nk and in the point i. Similarly we define $v^0 \in V_h$, the discretization of the initial data v_0.

The nonlinear algebraic system (4.4), (4.5) has a unique solution, according to Elliott and Ockendon [45], Ch.III. Denoting by $y_{h,k}$ the step function obtained starting from the solution of (4.4), (4.5), we show that the iterated limit $y_{h,k}$, with respect to $h,k \to 0$, is a weak solution of the two-phase Stefan problem, according to (1.7).

Theorem 4.1. We have

$$\lim_{k \to 0} \lim_{h \to 0} y_{h,k} = y$$

the weak solution of (3.2) - (3.5). The limit with respect to h is in the strong topology of $L^2(\Omega)$ and with respect to k is in the weak topology of $L^2(0,T,;H^1(\Omega))$.

Proof

We suppress the indices h,k. We put $v = y^{n+1}$ in (4.4) and we denote $w^n \in \gamma(y^n)$. We have

(4.6) $$1/k(y^{n+1} - y^n, y^{n+1})_h + 1/k(w^{n+1} - w^n, y^{n+1})_h +$$

$$+ |\text{grad } y^{n+1}|_H^2 = \int_{\partial\Omega} u^{n+1} y^{n+1} + (f^{n+1}, y^{n+1})_h.$$

Let $j : R \to]-\infty, +\infty]$ be the convex, lower semicontinuous function, such that $\gamma = \partial j$ and $j^* : R \to]-\infty, +\infty]$,

$$j^*(r) = \sup\{rp - j(p), p \in R\},$$

be the Fenchel conjugate of j. By the definition of the subdifferential, we have

$$(w^{n+1} - w^n, y^{n+1})_h = 1/3 \sum_{i \leq I} w_i (w_i^{n+1} - w_i^n) y_i^{n+1} \geq$$

$$\geq \frac{1}{3} \sum_{i \leq I} w_i [j^*(w_i^{n+1}) - j^*(w_i^n)].$$

We take the sum with respect to n and we infer

$$\sum_{n=0}^{p}(w^{n+1}-w^{n},y^{n+1})_h \geq \frac{1}{3}\sum_{n=0}^{p}\sum_{i\leq I} W_i[j^*(w_i^{n+1})-j^*(w_i^n)] =$$

$$= 1/3\sum_{i\leq I} W_i \sum_{n=0}^{p}[j^*(w_i^{n+1})-j^*(w_i^n)] = 1/3\sum_{i\leq I} W_i[j^*(w_i^{p+1}) -$$

$$j^*(w_i^0)] \geq C + 1/3\sum_{i\leq I} W_i[w_i^{p+1}y_i^{p+1} - j(y_i^{p+1})] \geq C + 1/3\sum_{i\leq I} W_i(-j(0)) \geq C.$$

Therefore

$$\sum_{n=0}^{p}(w^{n+1}-w^{n},y^{n+1})_h \geq C \tag{4.7}$$

and this inequality reflects the ciclical monotonicity of γ. By (4.6), (4.7) we deduce

$$1/k\sum_{n=0}^{p}(y^{n+1}-y^{n},\ y^{n+1})_h + C/k + \sum_{n=0}^{p}|\mathrm{grad}\ y^{n+1}|_H^2 \leq$$

$$\leq \sum_{n=0}^{p}\int_\Gamma u^{n+1}y^{n+1} + \sum_{n=0}^{p}(f^{n+1},y^{n+1})_h.$$

A simple computation gives

$$C/k + 1/2k|y^p|_h^2 - 1/2k|y^0|_h^2 + \sum_{n=0}^{p-1}|\mathrm{grad}\ y^{n+1}|_H^2 \leq$$

$$\leq C_1\sum_{n=0}^{p-1}\left\{|u^n|^2_{L^2(\partial\Omega)} + |f^n|_H^2\right\} + 1/2\sum_{n=0}^{p-1}|y^{n+1}|_V^2.$$

By (4.2) and the discrete Gronwall inequality, we get $\{y^n\}$ bounded in H with respect to h,k,n and

$$\sum_{n=0}^{m-1} k|y^{n+1}|_V^2 \leq C$$

for any h,k>0. Since β is a bounded operator, we also obtain $\{v^n\}$ bounded in H for all k,h,n.

Let k be fixed and $h\to 0$. On a subsequence, we have $y^n_{h,k}\to y^n_k$ strongly in $L^2(\Omega)$ and weakly in $H^1(\Omega)$. Because β is demiclosed, it yields $v^n_{h,k}\to v^n_k\in\beta(y^n_k)$ weakly in $L^2(\Omega)$.

Let us take any $v\in C^\infty(\bar\Omega)$ and let $v^h\in V_h$ denote the discretization of v, $v^h\to v$ for $h\to 0$, strongly in $H^1(\Omega)$. We replace v by v^h in (4.4) and, taking into account (4.3), we pass to the limit and obtain:

$$((v_k^{n+1}-v_k^n)/k,v) + \int_\Omega \mathrm{grad}\ y_k^{n+1}\mathrm{grad}\ v - \int_\Gamma u_k^{n+1}\ v = \int_\Omega f_k^{n+1}\ v,\quad v\in H^1(\Omega) \tag{4.8}$$

$$v_k^n\in\beta(y_k^n)\ \text{a.e.}\ \Omega, \tag{4.9}$$

where u_k^n, f_k^n are the limits with respect to $h\to 0$ of the discretizations associated with u,f. Obviously $v_k^0 = v_0$.

Since the solution of the semidiscrete problem (4.8), (4.9) is unique, the convergence is true without taking subsequences.

By the weak lower semicontinuity of the norm and the above estimates, we have $\{y_k^n\}$ bounded in H for any k,n and

$$\sum_{n=0}^{m} k|y_k^n|_V^2 \leq C, \quad k > 0. \tag{4.10}$$

The relation (4.6) and the definition of the dual norm shows that

$$|(v_k^{n+1} - v_k^n)/k|_{H^1(\Omega)^*} \leq C(|y_k^{n+1}|_{H^1(\Omega)} + |u_k^{n+1}|_{L^2(\partial\Omega)} + |f_k^{n+1}|_H)$$

and (4.10) implies

$$\sum_{n=0}^{m-1} k|(v_k^{n+1} - v_k^n)/k|_{H^1(\Omega)^*}^2 \leq C, \quad k > 0. \tag{4.11}$$

Then, the polygonal function defined from the vector $\{v_k^n\}$, denoted v_k, is bounded in $L^\infty(0,T;H)$ and $\{dv_k/dt\}$ is bounded in $L^2(0,T;H^1(\Omega)^*)$.

The Aubin compactness theorem gives $v_k \to \tilde{v}$ strongly in $L^2(0,T;H^1(\Omega)^*)$. By (4.10), we know that $y_k \to \tilde{y}$ weakly in $L^2(0,T;H^1(\Omega))$, where y_k is the polygonal function on [0,T], induced by the vector $\{y_k^n\}$. As the maximal monotone operators are demiclosed, (4.9) implies

$$\tilde{v} \in \beta(\tilde{y}) \qquad \text{a.e.}\, Q. \tag{4.12}$$

We show that $[\tilde{y},\tilde{v}]$ is a weak solution for the two-phase Stefan problem. We consider $z \in C^\infty(\bar{Q})$, $z(x,T) = 0$ and the discretization

$$z_{h,k}(t) = (z(nk))_I = z^n \in V_h, \text{ for } t \in]nk,(n+1)k]$$
$$z'_{h,k}(t) = (z^{n+1} - z^n)/k, \ t \in]nk,(n+1)k].$$

It is known that $z_{h,k} \to z$ strongly in $L^2(0,T;V)$, $z'_{h,k} \to \partial z/\partial t$ strongly in $L^2(Q)$.

We return to (4.4) and we put $v = z^n$. Summing by parts, we obtain

$$0 = k\sum_{n=0}^{m-1} ((v^{n+1} - v^n)/k, z^n)_h + \sum_{n=0}^{m-1} k\int_\Omega \operatorname{grad} y^{n+1} \operatorname{grad} z^n -$$
$$\sum_{n=0}^{m-1} k\int_{\partial\Omega} \dot{u}^{n+1} z^n - \sum_{n=0}^{m-1} k(f^{n+1}, z^n)_h,$$

$$(v^m, z^{m-1})_h - (v^0 z^0)_h + k\sum_{n=1}^{m-1} (-v^n, (z^n - z^{n-1})/k)_h +$$
$$k\sum_{n=0}^{m-1} \int_\Omega \operatorname{grad} y^{n+1} \operatorname{grad} z^n - k\sum_{n=0}^{m-1} \int_{\partial\Omega} u^{n+1} z^n - k\sum_{n=0}^{m-1} (f^{n+1}, z^n)_h = 0.$$

According to (4.3) and the established estimates, we can pass to the limit in this equality, first $h \to 0$, next $k \to 0$, and obtain the desired conclusion. Since the weak solution is unique, the convergence with respect to k is valid on the whole sequence, as with respect to h.

Theorem 4.2. We assume that $u \in W^{1,2}(0,T;L^2(\partial\Omega))$. Then $y_{h,k} \to y$ strongly in $L^2(Q)$, where y is the weak solution of the problem (3.2) - (3.5).

Proof

We put $v = y^{n+1} - y^n$ in (4.4):

$$1/k\,|y^{n+1} - y^n|_h^2 + \int_\Omega \operatorname{grad} y^{n+1}(\operatorname{grad} y^{n+1} - \operatorname{grad} y^n) \le$$

$$\le \int_{\partial\Omega} u^{n+1}(y^{n+1} - y^n) + (f^{n+1}, y^{n+1} - y^n)_h.$$

A simple calculation implies

$$1/2\,|\operatorname{grad} y^p|_H^2 - 1/2|\operatorname{grad} y^0|_H^2 + 1/2\sum_{n=0}^{p-1}|\operatorname{grad} y^{n+1} - \operatorname{grad} y^n|_H^2 +$$

$$+ 1/k\sum_{n=0}^{p-1}|y^{n+1} - y^n|_h^2 \le \sum_{n=0}^{p-1}\int_{\partial\Omega} u^{n+1}(y^{n+1} - y^n) + \sum_{n=0}^{p-1}(f^{n+1}, y^{n+1} - y^n)_h.$$

The right-hand side of this inequality may be estimated as follows

$$\sum_{n=0}^{p-1}\int_{\partial\Omega} u^{n+1}(y^{n+1} - y^n) = \int_{\partial\Omega} u^p y^p - \int_{\partial\Omega} u^0 y^0 + \sum_{n=1}^{p-1}\int_{\partial\Omega}(u^n - u^{n+1})y^n \le$$

$$\le C(1 + |\operatorname{grad} y^p|_H) + \sum_{n=1}^{p-1}\int_{\partial\Omega} k/2\,|(u^n - u^{n+1})/k|^2 + 1/2\sum_{n=1}^{p-1} k\,|y^n|_V^2 \le C(1 + |\operatorname{grad} y^p|_H),$$

by hypothesis and by (4.10). We continue with:

$$\sum_{n=0}^{p-1}(f^{n+1}, y^{n+1} - y^n)_h = \sum_{n=0}^{p-1}(k^{1/2} f^{n+1}, k^{-1/2}(y^{n+1} - y^n))_h \le$$

$$\le C\sum_{n=0}^{p-1} k\,|f^{n+1}|_H^2 + 1/2k\sum_{n=0}^{p-1}|y^{n+1} - y^n|_H^2.$$

Combining the above inequalities, we have

$$1/2\,|\operatorname{grad} y^p|_H^2 - 1/2\,|\operatorname{grad} y^0|_H^2 + 1/2k\sum_{n=0}^{p-1}|y^{n+1} - y^n|_H^2 \le C(1 + |\operatorname{grad} y^p|_H)$$

and we get $\{y^n\}$ bounded in V for any k,h,n and

$$\sum_{n=0}^{m-1} k\,|(y^{n+1} - y^n)/k|_h^2 \le C, \qquad\qquad h,k > 0.$$

We consider the interpolates

$y_{h,k}(t) = y^n$, $t \in]nk, (n+1)k]$,

$v_{h,k}(t) = v^n$, $t \in]nk, (n+1)k]$,

$\hat{y}_{h,k}(t) = ((n+1)k - t)/k\, y^n + (t - nk)/k\, y^{n+1}$, $t \in]nk, (n+1)k]$.

We have shown that $\{y_{h,k}\}$ is bounded in $L^\infty(0,T;H^1(\Omega))$, $\{v_{h,k}\}$ is bounded in $L^\infty(0,T;L^2(\Omega))$, $\{\hat{y}_{h,k}\}$ is bounded in $H^1(Q)$ and we deduce that $\hat{y}_{h,k} \to \tilde{y}$ strongly in $L^2(Q)$, $y_{h,k} \to \tilde{y}$ strongly in $L^2(Q)$, $v_{h,k} \to \tilde{v} \in \beta(\tilde{y})$ weakly* in $L^\infty(0,T;L^2(\Omega))$. The proof that $[\tilde{y},\tilde{v}]$ is a weak solution for (3.2) - (3.5) follows the same lines as in Thm.4.1. By the uniqueness of the weak solution, it yields that the above convergences are valid without taking subsequences.

Remark 4.3. The continuous dependence on the data of the variational solution in the two-phase Stefan problem, establishes a relationship between Thm.4.1 and Thm.4.2. The following result is a completion of Proposition 3.1.

Proposition 4.4. *Let $u_1, u_2 \in L^2(\Sigma)$ and y_1, y_2 be the corresponding variational solutions of (3.2) - (3.5). Then*

$$|y_1 - y_2|_{L^2(Q)} \le C|u_1 - u_2|_{L^2(\Sigma)},$$

where C is independent of u_1, u_2.

Proof.

Let $w_1 = \int_0^t y_1$, $w_2 = \int_0^t y_2$ satisfy (1.12), (1.13) (or (3.6) in the abstract variant), according to the definition of the variational solution.

We subtract the equations and multiply by $dw_1/dt - dw_2/dt$:

$$\int_0^t \int_\Omega |\partial w_1/\partial t - \partial w_2/\partial t|^2 - 1/2 \int_\Omega \Delta(w_1 - w_2)(w_1 - w_2) \le 0, \quad t \in [0,T]. \tag{4.13}$$

Integrating by parts in (4.13), we get

$$\int_0^t \int_\Omega |\partial w_1/\partial t - \partial w_2/\partial t|^2 + 1/2 \int_\Omega |\operatorname{grad}(w_1 - w_2)|^2 \le C|u_1 - u_2|_{L^2(\Sigma)} |w_1(t) - w_2(t)|_V$$

and the proof is finished.

Remark 4.5. Numerical examples and solutions of two-phase Stefan problems may be found in [40], [45], [117].

Let us return to the control problem. Corollary 3.9 and Remark 3.10 show that, in order to find a suboptimal solution for the problem (3.1) - (3.5) it is enough to deal with the regularized problem (3.11) - (3.14).

By discretization, we obtain the following finite dimensional optimization problem:

$$\text{Minimize } k/2 \sum_{n=1}^{m} (|y^n - y_d^n|_h^2 + |u^n|^2_{L^2(\partial\Omega)}), \tag{4.14}$$

for $u \in (L_h)^m$, $y \in (V_h)^m$ such that

$$((v^{n+1} - v^n)/k, v)_h + \int_\Omega \operatorname{grad} y^{n+1} \operatorname{grad} v - \int_{\partial\Omega} u^{n+1} v = (f^{n+1}, v)_h, \tag{4.15}$$

$$v_i^n = \beta^\varepsilon(y_i^n), \ i \le I, \ n \le m-1, \tag{4.16}$$

for all $v \in V_h$ and for $\varepsilon > 0$ fixed as in § 3.

Here $\{y_d^n\} \in (V_h)^m$ is a discretization of $y_d \in L^2(Q)$.

<u>Proposition 4.6. The problem</u> (4.14) - (4.16) <u>has at least one optimal pair</u> $[y_{h,k}, u_{h,k}]$.

Since $\varepsilon > 0$ is fixed, we omit to mention the dependence on ε.

<u>Proof.</u>

Let $\{u_p\} \subset (L_h)^m$ be a minimizing sequence. By (4.14), $\{u_p\}$ is bounded in $L^2(\partial\Omega)^m$. Since $\mathcal{T}_h$ and the corresponding discretization of $\partial\Omega$ are fixed and u_p^n are piecewise linear mappings and continuous on $\partial\Omega$ (in dimension 1), it follows that the "coordinates" $\{u_{p,j}^n\}$, $n \leq m$, $j \leq J$ are bounded with respect to p. On a subsequence, again denoted by u_p, we have

$$u_{p,j}^n \to \tilde{u}_j^n, \qquad p \to \infty. \tag{4.17}$$

By the usual estimates, we see that

$$y_{p,j}^n \to \tilde{y}_j^n, \qquad p \to \infty, \tag{4.18}$$

where $\{y_p\} \subset (V_h)^m$ is the solution of (4.15), (4.16) corresponding to $\{u_p\}$.

The structure of the cost functional (4.14) and (4.17), (4.18) imply that $[\tilde{y}, \tilde{u}] \in (V_h)^m \times (L_h)^m$ is an optimal pair, which we denote $[y_{h,k}, u_{h,k}]$.

Let $\theta : (L_h)^m \to (V_h)^m$ be the mapping $u \to y$ given by (4.15), (4.16).

<u>Proposition 4.7.</u> θ <u>is Gateaux differentiable and for any</u> $u, w \in (L_h)^m$, $r = \nabla\theta(u)w \in (V_h)^m$ <u>and satisfies</u>

$$1/k(\dot{\beta}^\varepsilon(y^{n+1})r^{n+1} - \dot{\beta}^\varepsilon(y^n)r^n, v)_h + \int_\Omega \operatorname{grad} r^{n+1} \operatorname{grad} v = \int_{\partial\Omega} w^{n+1} v, \quad \forall\ v \in V_h,\ n \leq m-1, \tag{4.19}$$

$$r^0 = 0. \tag{4.20}$$

<u>Proof.</u>

We have

$$\nabla\theta(u)w = \lim_{\lambda \to 0} (\theta(u + \lambda w) - \theta(u)) / \lambda \qquad \text{in } (V_h)^m.$$

We denote $y = \theta(u)$, $\tilde{y} = \theta(u + \lambda w)$ and subtract the corresponding equations

$$1/k((\tilde{v}^{n+1} - v^{n+1})/\lambda, v)_h + \int_\Omega \operatorname{grad} (\tilde{y}^{n+1} - y^{n+1})/\lambda . \operatorname{grad} v =$$

$$= \int_{\partial\Omega} w^{n+1} v + 1/k((\tilde{v}^n - v^n)/\lambda, v)_h, \ \forall\ v \in V_h.$$

We take $v = (\tilde{y}^{n+1} - y^{n+1})/\lambda$. A simple calculation, based on the properties of β^ε, implies

$$1/k\, |(\tilde{y}^{n+1} - y^{n+1})/\lambda|_H^2 + |\operatorname{grad}(\tilde{y}^{n+1} - y^{n+1})/\lambda|_H^2 \leq C\, |(\tilde{y}^{n+1} - y^{n+1})/\lambda|_V +$$

$$+ 1/k\varepsilon\, |(\tilde{y}^n - y^n)/\lambda|_h . |(\tilde{y}^{n+1} - y^{n+1})/\lambda|_h.$$

Since m is fixed, we have

$|(\tilde{y}^n - y^n)/\lambda|_V$ bounded for $\lambda > 0$, $n \leq m$,

$|(\tilde{v}^n - v^n)/\lambda|_H$ bounded for $\lambda > 0$, $n \leq m$.

Consequently $\tilde{y}^n \to y^n$ in $H^1(\Omega)$ as $\lambda \to 0$ and, moreover, $\lim_{\lambda \to 0} (\tilde{y}^n - y^n)/\lambda = r$ exists in the strong topology of $L^2(\Omega)$, on a subsequence. Taking further subsequences, as in Proposition 4.6, we deduce that $r \in (V_h)^m$ and we have the coordinatewise convergence

$$(\tilde{y}_i^n - y_i^n)/\lambda \to r_i, \qquad i \leq I.$$

We can pass to the limit. The first term gives:

$$((\tilde{v}^n - v^n)/\lambda, v)_h = 1/3 \sum_{i \leq I} w_i (\beta^\varepsilon(\tilde{y}_i^n) - \beta^\varepsilon(y_i^n)/(\tilde{y}_i^n - y_i^n).$$

$$(\tilde{y}_i^n - y_i^n)/\lambda . v_i \to 1/3 \sum_{i \leq I} w_i \dot{\beta}^\varepsilon(y_i^n) r_i^n v_i = (\dot{\beta}^\varepsilon(y^n) r^n, v)_h,$$

which finishes the proof.

Let $J_{h,k} : (L_h)^m \to R$ be the cost functional (4.14). Then $u_{h,k}$ is a minimum point for $J_{h,k}$ and the Gateaux differential vanishes:

(4.21) $$\nabla J_{h,k}(u_{h,k}) w = 0, \qquad w \in (L_h)^m.$$

We have

$$\lim_{\lambda \to 0} (J_{h,k}(u + \lambda w) - J_{h,k}(u))/\lambda = k \sum_{n=1}^{m} ((y^n - y_d^n, r^n)_h + (u^n, w^n)_{L^2(\partial\Omega)}) =$$

$$= k \sum_{n=1}^{m} (([\nabla\theta(u)^*)(y - y_d)]^{n-1}, w^n)_{L^2(\partial\Omega)} + (u^n, w^n)_{L^2(\partial\Omega)}).$$

We define the adjoint state $p_{n,k} \in (V_h)^m$ by

(4.22) $$(\dot{\beta}^\varepsilon(y_{h,k}^n)(p^n - p^{n+1})/k, v)_h + \int_\Omega \operatorname{grad} p^n . \operatorname{grad} v =$$

$$= -(y_{h,k}^{n+1} - y_d^{n+1}, v)_h, \quad \forall\ v \in V_h,\ n \leq m-1,$$

(4.23) $$p^m = 0.$$

This is a linear system in implicit form and it has a unique solution $p_{h,k} \in (V_h)^m$.

<u>Proposition 4.8. The relation</u> (4.21) <u>is equivalent with</u>

(4.24) $$u_{h,k}^n = p_{h,k}^{n-1}|_{\partial\Omega}, \quad 1 \leq n \leq m.$$

Proof

We put in (4.19) $v = p^n_{h,k}$ and we omit the indices h,k:

$$(4.25)\qquad \sum_{n=0}^{m-1} ((\dot\beta^\varepsilon(y^{n+1})r^{n+1} - \dot\beta^\varepsilon(y^n)r^n)/k,\ p^n)_h +$$

$$+ \sum_{n=0}^{m-1} \int_\Omega \operatorname{grad} r^{n+1} \operatorname{grad} p^n = \sum_{n=0}^{m-1} \int_{\partial\Omega} w^{n+1} p^n.$$

Summing by parts in the first term, we get:

$$(4.26)\qquad \sum_{n=0}^{m-1} \int_{\partial\Omega} w^{n+1} p^n = \sum_{n=0}^{m-1} \int_\Omega \operatorname{grad} r^{n+1} \operatorname{grad} p^n + \sum_{n=1}^{m} (\dot\beta^\varepsilon(y^n)$$

$$(p^{n-1} - p^n)/k, r^n)_h = \sum_{n=0}^{m-1} \int_\Omega \operatorname{grad} r^{n+1} \operatorname{grad} p^n + \sum_{n=0}^{m-1} (\dot\beta^\varepsilon(y^{n+1})(p^n - p^{n+1})/k, r^{n+1})_h =$$

$$= -\sum_{n=0}^{m-1} (y^{n+1} - y_d^{n+1}, r^{n+1})_h = -\sum_{n=1}^{m} (y^n - y_d^n, r^n)_h,$$

according to (4.22). From (4.21), we have

$$\sum_{n=1}^{m} (u^n, w^n)_{L^2(\partial\Omega)} = -\sum_{n=1}^{m} (y^n - y_d^n, r^n)_h.$$

Combining with (4.25), (4.26), we infer

$$\sum_{n=1}^{m} \int_{\partial\Omega} u^n w^n = \sum_{n=0}^{m-1} \int_{\partial\Omega} w^{n+1} p^n, \qquad \forall\ w \in (L_h)^m$$

and (4.24) is proved.

<u>Theorem 4.9. We denote</u> J_ε <u>the functional defined by</u> (3.11) - (3.14) <u>and let</u> β <u>be strongly maximal monotone and sublinear. Then</u>

$$(4.27)\qquad \lim_{k\to 0} \lim_{h\to 0} J_\varepsilon(y_{h,k}, u_{h,k}) = J_\varepsilon(y_\varepsilon, u_\varepsilon),$$

<u>that is</u> $u_{h,k}$ <u>is a minimizing sequence for the problem</u> (3.11) - (3.14).

Proof

For any $h,k > 0$, we obtain

$$(4.28)\qquad k/2 \sum_{n=0}^{m-1} (|y^n_{h,k} - y^n_d|^2_h + |u^n_{h,k}|^2_{L^2(\partial\Omega)}) \le$$

$$\le k/2 \sum_{n=0}^{m-1} (|y^n - y^n_d|^2_h + |u^n|^2_{L^2(\partial\Omega)}),$$

for any $u \in (L_h)^m$, $y = \theta(u) \in (V_h)^m$.

Let $u \in W^{1,2}(0,T;L^2(\partial\Omega))$ and let $u^{h,k} \in (L_h)^m$ be the discretization of u and $y^{h,k} = \theta(u^{h,k})$. By (4.28) we deduce that $J_{h,k}(y_{h,k}, u_{h,k})$ is bounded with respect to $h,k > 0$.

The same estimates as in the proof of Thm.4.1 imply that $\{y^n_{h,k}\}$ is bounded in H, for any n,h,k and

$$\sum_{n=0}^{m-1} k|y^n_{h,k}|^2_V \leq C, \; h,k > 0. \tag{4.29}$$

Let $k > 0$ be fixed. By taking convenient subsequences, we have $u^n_{h,k} \to u^n_k$ weakly in $L^2(\partial\Omega)$, $y^n_{h,k} \to y^n_k$ strongly in $L^2(\Omega)$, for $h \to 0$.

Since β^ε is Lipschitzian, it yields that

$$v^n_{h,k} \to v^n_k = \beta^\varepsilon(y^n_k) \text{ strongly in } L^2(\Omega).$$

Obviously u^n_k, y^n_k, v^n_k satisfy (4.8), (4.9) with β replaced by β^ε. In a similar way as in Thm.4.1, we obtain the estimates

$$\sum_{n=0}^{m-1} (k|u^n_k|^2_{L^2(\partial\Omega)} + k|y^n_k|^2_V) + |y^n_k|^2_H \leq C, \tag{4.30}$$

$$\sum_{n=0}^{m-1} k|(v^{n+1}_k - v^n_k)/k|^2_{V^*} \leq C \tag{4.31}$$

for all $k > 0$, $n \leq m$.

Moreover, we can assume that $u_k \to \tilde{u}$ weakly in $L^2(\Sigma)$, $y_k \to \tilde{y}$ weakly in $L^2(0,T;H^1(\Omega))$, $v_k \to \tilde{v}$ strongly in $L^2(0,T;H^1(\Omega)^*)$ and $[\tilde{y},\tilde{v}]$ is a weak solution of the equation (3.2) - (3.5) corresponding to $\tilde{u}$ (and β is replaced by β^ε).

Due to (4.28) and to Proposition 4.4 it yields that $[\tilde{y},\tilde{u}]$ is an optimal pair for the problem (3.11)-(3.14) and we denote it $[y_\varepsilon, u_\varepsilon]$.

Now, we use the adjoint system. We take $v = p^n_{h,k} - p^{n+1}_{h,k}$ in (4.22):

$$k|(p^n - p^{n+1})/k|^2_h + \int_\Omega \operatorname{grad} p^n \operatorname{grad}(p^n - p^{n+1}) = -(y^n - y^n_d, p^n - p^{n+1})_h.$$

Summing with respect to n, after a short calculation, we see that $\{p^n\}$ is bounded in V for all n,h,k and

$$\sum_{n=0}^{m-1} k|(p^n - p^{n+1})/k|_h \leq C, \qquad \forall \; h,k > 0.$$

Let k be fixed. On a subsequence (which may be the same as above) we have

$$\lim_{h\to 0} p^n_{h,k} = p^n_k \text{ strongly in } H^{3/4}(\Omega).$$

The trace theorem and (4.24) give the strong convergence $u^n_{h,k} \to u^n_k$ in $L^2(\partial\Omega)$.

Since $\{p^n_k\}$ satisfies similar estimates, the Aubin theorem yields $\lim_{k\to 0} p^n_k = p$ strongly in $L^2(0,T;H^{3/4}(\Omega))$. So, on a subsequence, we have the iterated limit

(4.32) $\lim_{k\to 0} \lim_{h\to 0} u_{h,k} = u_\varepsilon$

strongly in $L^2(\Sigma)$.

We underline that, for $\{p_{h,k}\}$, it isn't necessary to consider the iterated limit, but for $\{u_{h,k}\}$ it is necessary, taking into account the convergence properties of $\{y_{h,k}\}$.

Now, Proposition 4.4 gives

$$\lim_{k\to 0} \lim_{h\to 0} \theta_\varepsilon(u_{h,k}) = y_\varepsilon$$

strongly in $L^2(Q)$. Here θ_ε is the mapping $u \to y$ defined by (3.12) - (3.14).

Due to the continuity of the functional (3.11), the proof is finished.

Remark 4.10. Corollary 3.9 and Thm.4.9 show that the problem (3.1) - (3.5) may be reduced to the finite dimensional problem (4.14) - (4.16).

We close this section by indicating a general gradient algorithm for the computation of $u_{h,k}$ in the problem (4.14) - (4.16):

Step 1 - we choose u_0 and we put $s = 0$;
Step 2 - we compute y_s, the solution of (4.15), (4.16), corresponding to u_s;
Step 3 - we test the pair $[y_s, u_s]$ if it is satisfactory; if YES, then STOP; otherwise GO TO step 4;
Step 4 - we compute p_s by (4.22), (4.23);
Step 5 - we compute u_{s+1} by the formula $u_{s+1} = u_s - \rho_s(u_s - p_s|_\Sigma)$ where ρ_s is a real parameter;
Step 6 - we put $s \to s+1$ and GO TO step 2.

The convergence test considered in step 3 is that $|J_{h,k}(u_s) - J_{h,k}(u_{s+1})|$ or $|u_s - p_s|_{L^2(\Sigma)}$ is smaller then a given parameter. In step 5, ρ_s may be chossen by a line search.

It is known that, without convexity properties, such algorithms may not converge to a minimum point of $J_{h,k}$ (see [36], [95]). Because $J_{h,k}$ isn't convex, the next result stresses the descent property of the algorithm.

Proposition 4.11. i) Let $h,k > 0$ be fixed. The sequence $J_{h,k}(u_s)$ is convergent when $s \to \infty$.

ii) Assume that the first iteration u_0 is sufficiently regular and let $\tilde{u}_{h,k}$ be the value for $u_{h,k}$ given by the algorithm. The sequence $J_{h,k}(\tilde{u}_{h,k})$ is bounded with respect to $h,k > 0$ and any limit point $\tilde{J}$ satisfies

$$J_\varepsilon(y_\varepsilon, u_\varepsilon) \le \tilde{J} \le J_\varepsilon(y^0, u_0).$$

Here $y^0 = \theta_\varepsilon(u_0)$ is given by (3.12) - (3.14).

Proof.

i) The sequence is decreasing and bounded from below.

ii) We assume that u_0 is sufficiently regular, such that $u_0^{j,k}$ (the discretization of u_0) converges to u_0 strongly in $L^2(\Sigma)$, for $h,k\to 0$. We have

$$J_{h,k}(y_{h,k},u_{h,k})\leq J_{h,k}(\tilde{y}_{h,k},\tilde{u}_{h,k})\leq J_{h,k}(y^o_{h,k},u_0^{h,k}),$$

where $\tilde{y}_{h,k} = \theta_\varepsilon((\tilde{u}_{h;,k}),\ y^o_{h,k} = \theta_\varepsilon(u_0^{h,k})$.

By the properties of $u_0^{h,k}$ and by Thm.4.9 we can pass to the limit and obtain the result.

Remark 4.12. The significance of Proposition 4.11 is that, in real problems, we don't look for the optimal performance since $J_{h,k}(\tilde{u}_{h,k})$ may be different from $J_{h,k}(u_{h,k})$. We start the algorithm with a control u_0 already used in practice and we improve the performance given by u_0. If u_0 is not sufficiently regular, we may replace it with a regular approximation $\bar{u}_0$, due to Proposition 4.4.

Remark 4.13. In our attempt to justify mathematically the numerical computations, we have studied $\tilde{u}_{h,k}$, the computed values. It would be useful to prove a similar statement for the sequence $J_\varepsilon(u_{h,k})$.

Remark 4.14. Numerical results for control problems governed by two-phase Stefan problems may be found in the papers [91], [93], [22], [109], [118].

REFERENCES

1) **E. Asplund** - "Averaged norms", Israel J. Math. 5(1967).

2) **J.P. Aubin** - "Un theoreme de compacite, C.R.Acad, Sci.Paris, 256(1963).

3) **J.P.Aubin** - "Methodes explicites de l'optimisation", Dunod, Paris (1982).

4) **W.F.Ames** - "Nonlinear partial differential equations in engineering", Academic Press, New York - London (1965).

5) **S.Agmon, A.Douglis, L.Nirenberg** - "Estimates near boundary for solutions of elliptic partial differential equations satisfying general boundary conditions", Comm.Pure, Appl. Math. 12(1959).

6) **L.Amerio** - "Unilateral problems for hyperbolic equations in two variables". In: "Free Boundary Problems", Proceedings, Ist.Naz.Alta Matematica "Francesco Severi", Rome (1980).

7) **L.Amerio, G.Prouse** - "Study of the motion of a string vibrating against an obstacle", Rend. Mat. 2(1975).

8) **D.R.Atthey** - "A finite difference scheme for melting problems", J. Inst. Maths. Applics. 13(1974).

9) **V.Arnautu** - "Characterization and approximation of a class of nonconvex distributed control problems, Mathematica 2(1980).

10) **V.Arnautu** - "Approximation of optimal distributed control problems governed by variational inequalities", Numer.Math.38(1982).

11) **M. Badii, G.F. Webb** - "Nonlinear nonautonomous functional differential equations in L^p spaces", Nonlin. Anal. TMA 5(1981).

12) **V.Barbu** - "Nonlinear semigroups and differential equations in Banach spaces", Noordhoff, Leyden, the Netherlands (1976).

13) **V.Barbu** - "Optimal control of variational inequalities", Research notes in mathematics 100, Pitman, Boston - London - Melbourne (1984).

14) **V.Barbu, Th.Precupanu** - "Convexity and optimization in Banach spaces", D.Reidel, Dordrecht - Boston - Lancaster (1986).

15) **V.Barbu** - "Boundary control problems with nonlinear state equations", SIAM J.Control and Optimiz.20 (1982).

16) V.Barbu - "Necessary conditions for multiple integral problem in the calculus of variations", Math.Ann. 260(2) (1982).

17) **V.Barbu** - "Optimal control for free boundary problems", Conf.del Sem. di Mat. Univ.Bari, 206(1985).

18) **V.Barbu, V.Arnautu** - "Optimal control of the free boundary in a two-phase Stefan problem", Preprint Series in Mathematics 11, INCREST, Bucuresti (1985).

19) **V. Barbu, D. Tiba** - "Optimal control of abstract variational inequalities" in "Control fo distributed parameter systems", M. Amouroux and A. El Jai Eds., Pergamon Press, Oxford (1990).

20) **H.Brezis** - "Problemes unilateraux", J.Math.pures et appl. 51(1972).

21) **H.Brezis** - "Operateurs maximaux monotones et semi-groupes de contractions dans les espaces de Hilbert", North Holland, Amsterdam - London (1973).

22) **J.F.Bonnans** - These, Univ. de Technologie de Compiegne (1982).

23) **J.F. Bonnans, E. Casas** - "A principle of Pontryagin for the optimal control of semilinear elliptic systems", to appear in J. of Diff. Eq. (1990).

24) **J.F. Bonnans, D. Tiba** - "Pontryagin's principle in the control of semilinear elliptic vaiational inequalities", to appear in Appl. Math. and Optimiz. (1990).

25) **C.Baiocchi** - "Su un problema a frontiera libera conesso a questioni di idraulica", Ann. Mat. Pure ed Applicata 92(1972).

26) **A.Bermudez, C.Saguez** - "Optimal control of variational inequalities", Proceedings of the 23rd C.D.C., Las Vegas (1984).

27) **E. Di Benedetto, R.E.Showalter** - "Implicit degenerate evolution equations and applications", SIAM J. Math.Anal., 12(1981).

28) **E.Di Benedetto** - "Continuity of weak solutions to certain singular parabolic equations", Ann. di Mat. Pura, Appl. CXXX(1982).

29) **M.Brokate** - "Necessary optimality conditions for the control of the semilinear hyperbolic boundary value problems", Preprint nr.54, Univ. Augsburg (1985).

30) **M. Brokate, J. Sprekels** - "Optimal control of thermomechanical phase transitions in shape memory alloys: necessary conditions of optimality", Report 199, Univ. GHS Essen (1990).

31) **F.H.Clarke** - "Generalized gradients and applications", Trans. Amer. Math. Soc. 205 (1975).

32) **F.H. Clarke** - "Optimization and nonsmooth analysis", John Wiley and Sons, New York (1983).

33) **C.Citrini** - "The vibrating string with pointshaped obstacle". In: "Free Boundary Problems", Proceedings, Ist.Naz. Alta Matematica "Francesco Severi", Rome (1980).

34) **E.Casas** - Thesis, Universidad de Santiago de Compostela (1982).

35) **J.Cea, E.J.Haug** (ed.) - "Optimization of distributed parameter structures", Sijthoff and Noordhoff (1981).

36) **J.Cea** - "Optimisation, theorie et algorithmes", Dunod, Paris (1971).

37) **L.A.Caffarelli, L.C.Evans** - "Continuity of the temperature in the two-phase Stefan problem", Arch.Rat.Mech.Anal.81(1983).

38) **L.A.Caffarelli, A.Friedman** - "Continuity of the temperature in the Stefan problem", Indiana U.Math.J.28(1979).

39) **Ph.G.Ciarlet** - "The finite element method for elliptic problems", North Holland, Amsterdam (1978).

40) **J.F.Ciavaldini** - "Analyse numerique d'un probleme de Stefan a deux phases par un methode d'elements finis", SIAM J.Numer.Anal.12(1975).

41) **G.Duvaut** - Resolution d'un probleme de Stefan", C.R.Acad.Sci.Paris 276(1973).

42) **J.Ekeland, R.Temam** - "Analyse convexe et problemes variationnels", Dunod, Gauthier-Villars, Paris (1974).

43) **I.Ekeland** - "Sur les problems variationnels", C.R.A.S. Paris 275(1972), p. 1057-1059.

44) **I. Ekeland** - "Nonconvex minimization problems", Bull. Amer. Math. Soc. 1(N.S.), (1979), p. 447-474.

45) **C.M.Elliott, J.R.Ockendon** - "Weak and variational methods for moving boundary problems", Research Notes in Mathematics 59, Pitman, Boston - London - Melbourne (1982).

46) **L.C.Evans** - "Differentiability of a nonlinear semigroup in L^1" J. Math. Anal. Appl. 60(1977).

47) **A.Friedman** - "Variational principles and free boundary problems", J.Willey and Sons (1982).

48) **A.Friedman, L.S.Jiang** - "Nonlinear optimal control in heat conduction", SIAM J.Control and Optimiz. 21(1983).

49) **A.Friedman** - "Nonlinear optimal control for parabolic equations" - SIAM J.Control and Optimiz. 22(1984).

50) **A.Friedman** - "Optimal control for parabolic variational inequalities" - SIAM J.Control and Optimiz. 24(1986).

51) **A.Friedman, D.Yaniro** - "Optimal control for the dam problem", Appl. Math. Optim. 13(1985).

52) **G.Fix, G.Strang** - An analysis of the finite element method", Prentice - Hall (1973).

53) **M.Fremond** - Autumn course on applications of analysis to mechanics", I.C.T.P. Trieste (1976).

54) **P. Grisvard** - "Elliptic problems in nonsmooth domains", Pitman, London (1985).

55) **E.Hewitt, K.Stromberg** - "Real and abstract analysis", Springer Verlag, Berlin-Heidelberg-New York (1965).

56) **Zheng - Xu He** - "State constrained control problems governed by variational inequalities", SIAM J.Control and Optimiz. 25(1987).

57) **Zheng - Xu He, Gh.Morosanu** - "Optimal control of biharmonic variational inequalities", Preprint Series in Mathematics, Univ."Al.I.Cuza", Iasi (1985).

58) **I.Hlavacek, I.Bock, J.Lovisek** - "Optimal control of a variational inequality with applications to structural analysis. I.Optimal design of a beam with unilateral supports", Appl.Math.Optimiz. 11(1984).

59) **J.Haslinger, P.Neittaanmaki** - "Finite element approximation for optimal shape design. Theory and applications", John Wiley, New York (1988).

60) **K.H.Hoffmann, J.Sprekels** - "Real time control in a free boundary problem connected with the continuous casting of steel", In: "Optimal control of partial differential equations", Proceedings, Birkhauser Verlag, Basel - Boston - Stuttgart (1984).

61) **J.W.Jerome, M.E.Rose** - "Error estimates for the multidimensional two - phase Stefan problem", Math.Comput. 39(1982).

62) **D.Kinderlehrer, G.Stampacchia** - "An introduction to variational inequalities and their applications", Academic Press (1980).

63) **E.Krauss** - "A representation of maximal monotone operators by saddle functions" Rev. Roum. Math.Pures, Appl.10(1985).

64) **S.L.Kamenomostskaja** - "On Stefan's problem" Mat.Sb.53(1961), (in Russian).

65) **S.L.Kamenomostskaja** - "On Stefan's problem", Nauen.Dokl. Vyss. Skoly 1(1958) (in Russian).

66) **O.A. Ladyshenskaya** - "The boundary value problems of mathematical physics", Springer-Verlag, Berlin (1985).

67) **O.A. Ladyzhenskaya, V.A. Solonnikov, N.N. Uralceva** - "Linear and quasilinear equations of parabolic type", Translations AMS, Providence, RI (1968).

68) **J.L.Lions** - "Controle optimal des systemes gouvernes par des equations aux derivees partielles", Dunod, Paris (1968).

69) **J.L.Lions** - "Quelques methodes de resolution des problemes aux limites non lineaires", Dunod, Paris (1969).

70) **J.L.Lions, E.Magenes** - "Problemes aux limites non homogenes et applications", Dunod, Paris (1968).

71) **J.L.Lions** - "Various Topics in the theory of optimal control of distributed systems", Lect. Notes in Economics Sci. 105, Springer-Verlag, Berlin (1974).

72) **J.L.Lions** - "Controle des systemes distribues singuliers", Dunod, Paris (1983).

73) **J.L.Lions, G.Stampacchia** - "Variational inequalities", C.P.A.M. XX (1967).

74) **F.Mignot** - "Controle dans les inequations variationnelles elliptiques", J.Funct.Anal.22(1976).

75) **F.Mignot, J.P.Puel** - "Optimal control in some variational inequalities", SIAM J. Control and Optimiz., 22(3) (1984).

76) **F.Mignot, J.P.Puel** - "Controle optimal d'un systeme gouverne par une inequation variationnelle parabolique", C.R.A.S. Paris 298 (1984).

77) **V.Mikhailov** - "Equations aux derivees partielles", Moscou, Editions Mir (1980).

78) **A.Meirmanov** - "On the classical solution of the multidimensional Stefan problem for quasilinear parabolic equations", Matem.Sbornik 112 (1980).

79) **Z.Meike** - "Existence for an evolution equation with a nonmonotone continuous nonlinearity", Nonlin.Anal.TM A 5(1981).

80) **M.Nicolescu** - "Analiza matematica", vol.II, Editura Tehnica, Bucuresti (1958), (in Romanian).

81) **J.Naumann** - "Parabolische Variationsungleichnungen", Teubner Texte, Leipzig (1985).

82) **M.Niezgodka** - "Stefan - like problems" In: "Free boundary problems - Theory and applications", Proceedings, A.Fasano, M.Primicerio Eds., Research Notes in Mathematics 78, 79, Pitman, Boston - London - Melbourne (1982).

83) **M.Niezgodka, I.Pawlow** - "A generalized Stefan problem in several space variables", Appl.Math.Optim.9(1983).

84) **M.Niezgodka, I.Pawlow** - "Optimal control for parabolic systems with free boundaries", In: "Optimization Techniques", K.Iracki, K.Malanowski, S.Walukiewicz, Eds., Lect. Notes Control Inform.Sci.22, Springer Verlag, Berlin (1980).

85) **M.Niezgodka** - "Modelling and approximation of free optimization", K.Malanowski, K.Mizukami Eds., Warsawa (1985).

86) **M.Niezgodka, I.Pawlow, A.Visintin** - "On multi - phase Stefan type problems with nonlinear flux at the boundary in several space variables", Preprint 293, Ist.di Anal. Numer., Pavia (1981).

87) **J.T.Oden** - "Finite elements of nonlinear continua", Mc.Graw - Hill (1972).

88) **O.Pirronneau** - "Optimal shape design for elliptic systems", Springer Verlag, Berlin-Heidelberg-New York-Tokyo (1984).

89) **M.Primicerio** - "Mushy region in phase change problems", In: "Applied nonlinear functional analysis", R.Gorenflo, K.H.Hoffmann Eds., P.Lang Verlag, Frankfurt Main (1983).

90) **I.Pawlow** - "Analysis and control of evolution multi-phase problems with free boundaries", Ossolineum, Warszawa (1987).

91) **I.Pawlow** - "Variational inequality formulation and optimal control of nonlinear evolution systems governed by free boundary problems", In: "Applied nonlinear functional analysis", R.Gorenflo, K.H.Hoffmann, Eds., P.Lang Verlag, Frankfurt Main (1983).

92) **I.Pawlow** - "A variational inequality approach to generalized two-phase Stefan problem in several space variables", Ann. di Mat.Pura, Appl.CXXXI (1982).

93) **I.Pawlow** - "Optimal control of nonlinear evolution problems with applications to processes involving free boundaries". In: "Constructive aspects of optimization", K.Malanowski, K.Mizukami, Eds., Warsawa (1985).

94) **I.Pawlow** - "Approximation of an evolution variational inequality arising from free boundary problems", In: "Optimal control of partial differential equations", K.H.Hoffmann, W.Krabs, Eds., Birkhauser, Basel-Boston-Stuttgart (1984).

95) **B.N.Pshenichny, Yu.M.Danilin** - "Numerical methods in extremal problems", Mir Publishers, Moscow (1978).

96) **L.S. Pontryagin, V.G. Boltianski, R.V. Gamkrelidze, E.E. Mischenko** - "The mathematical theory of optimal processes", New York, Interscience Publishers (1962).

97) **R. Phelps** - "Convex functions, monotone operators and differentiability", LNM 1364, Springer Verlag, Berlin (1989).

98) **R.T. Rockafellar** - "On the maximal monotonicity of subdifferential mappings", Pacific J. Math. 33 (1970).

99) **R.T.Rockafellar** - "Convex Analysis", Princeton Univ. Press (1970).

100) **L.I.Rubenstein** - "The Stefan problem", A.M.S., Providence, Rhode Island (1971).

101) **L.I.Rubenstein** - "Application of the integral equation technique to the solution of several Stefan problems", In: "Free Boundary Problems", Proceedings, Ist.Naz. Alta Matematica "Francesco Severi", Rome (1980).

102) **T.Roubicek** - Optimal control of a Stefan problem with state-space constraints. Numerical approximation", Numerische Mathematik (1987).

103) **V.A. Solonnikov** - "A priori estimates for equations of second order parabolic type" Tr.Mat.Inst.Steklov, 10(1964).

104) **M.Schatzman** - These, Univ. "Pierre et Marie Curie", Paris (1979).

105) **Shi Shuzhong** - "Optimal control of strongly monotone variational inequalities", SIAM

J. Control and Optimiz., vol. 26 (1988), p. 274-291.

106) **J.Sokolowski, A.Myslinski** - "Nondifferentiable optimization problems for elliptic systems", SIAM J.Control and Optimiz.23(1985).

107) **J.Stefan** - "Uber einige Probleme der Theorie der Warmleitung", S. - B.Wien, Akad.Mat.Natur. 98(1889).

108) **C.Saguez, A.Bermudez** - "Modelisation et simulation d'un alliage a n composants", In: Analysis and Optimization of Systems", Proceedings, A.Bensoussan, J.L.Lions, Eds., Lect. Notes Control Inform.Sci.44, Soringer Verlag, Berlin (1982).

109) **C.Saguez** - These, Univ.de Technologie de Compiegne (1980).

110) **R.A.Tapia** - "The differentiation and integration of nonlinear operators", In: "Nonlinear functional analysis and applications", L.B.Rall Ed., Academic Press, New York-London (1971).

111) **D.A.Tarzia** - "Etude de l'inequation variationnelle proposee par Duvaut pour le probleme de Stefan a deux phases", Bollettino U.M.I. 1-B (1982).

112) **D.Tiba** - "Subdifferentials of composed functions and applications in optimal control", An.St.Univ. "Al.I.Cuza", Iasi 23(1977).

113) **D.Tiba** - "Regularization of saddle functions", Bollettino U.M.I. 17-A (1980).

114) **D.Tiba, Z.Meike** - "Optimal control for a Stefan problem", In: "Analysis and Optimization of Systems", Proceedings, A.Bensoussan, J.L.Lions, Eds., Lecture Notes in Control and Inform. Sci. 44, Springer Verlag, Berlin (1982).

115) **D.Tiba** - "Boundary control for a Stefan problem", In: "Optimal control of partial differential equations", K.-H.Hoffmann, W.Krabs, Eds., Birkhauser, Basel-Boston-Stuttgart (1984).

116) **D. Tiba, M. Tiba** - "Approximation for control problems with pointwise state constraints", ISNM91, Birkhauser Verlag, Basel, p. 379-390 (1989).

117) **D.Tiba, M.Tiba** - "Regularity of the boundary data and the convergence of the finite element discretization in two-phase Stefan problems", Int.J.Engng.Sci. 22(1984).

118) **D.Tiba, P.Neittaanmaki** - "On the finite element approximation of the boundary control for two-phase Stefan problems", In: "Analysis and Optimization of Systems", A.Bensoussan, J.L.Lions, Eds., Lect.Notes in Control and Inform.Sci.62, Springer Verlag, Berlin (1984).

119) **D.Tiba** - "Criterii de optimalitate pentru probleme convexe de control. Conditii necesare in probleme cu ecuatie neliniara de stare", Studii si Cercetari Matematice 6(1984), (in Romanian).

120) **D.Tiba** - "Quelques remarques sur le controle de la corde vibrante avec obstacle", C.R. Acad. Sci. Paris 299(1984).

121) **D.Tiba** - "Some remarks on the control of the vibrating string with obstacle", Rev. Roum. Math.Pures, Appl. XXIX(1984).

122) **D.Tiba** - Teza, Univ. "Al.I.Cuza" Iasi (1982), (in Romanian).

123) **D.Tiba, V.Komornik** - "On the control of strongly nonlinear systems", Rapport de Recherche 352, INRIA, Paris (1984).

124) **D.Tiba, V.Komornik** - Controles de systemes fortement non lineaires", C.R.Acad. Sci. Paris, 300(1985).

125) **D.Tiba** - "Optimality conditions for distributed control problems with nonlinear state equation", SIAM J.Control and Optimiz" 23(1985).

126) **D.Tiba, E.Krauss** - "Regularization of saddle functions and the Yosida approximation of monotone operators", An.St.Univ. "Al.I.Cuza" Iasi, XXXI (1985).

127) **D.Tiba, E.Krauss** - "Anisotropic regularization of saddle functions", Preprint Series in Mathematics 28, INCREST, Bucuresti (1985).

128) **D.Tiba** - "Optimal control of hyperbolic variational inequalities" In: "Nondifferentiable optimization, motivations and applications", V.F.Demyanov, D.Pallaschke, Eds., Lect.Notes Econ. and Math. Systems 255, Springer Verlag, Berlin(1985).

129) **D.Tiba** - "Une approche par inequations variationnelles pour les problemes de controle avec contraintes" C.R.Acad.Sci.Paris, 302(1986).

130) **D.Tiba, P.Neittannmaki** - "A variational inequality approach in constrained control problems", Appl.Math.Optimiz., (1988).

131) **D.Tiba, P.Neittaanmaki, J.Haslinger** - "A variational inequality approach in optimal design problems", In: "Optimal control of partial differential equations II", K.-H.Hoffmann, W.Krabs, Eds., ISNM 78, Birkhauser Verlag, Basel-Boston-Stuttgart (1987).

132) **D.Tiba, J.F.Bonnans** - "Equivalent control problems and applications", in: "Control problems for systems described by partial differential equations", Lasiecka, R. Tuggiani, Eds., LNCIS 97, Springer Verlag (1987).

133) **D.Tiba** - "Optimal control for second order semilinear hyperbolic equations", Control, Theory and Advanced Technology, vol.3, nr.1, (1987).

134) **F.Troltzsch** - "Optimality conditions for parabolic control problems and applications", Teubner Texte, Leipzig (1984).

135) **M.Ughi** - "A melting problem with a mushy region: qualitative properties", IMA J.Appl.Math.33(1984).

136) **A.Visintin** - "Strong convergence results related to strict convexity", Comm. in Part.Diff.Eq.9(1984).

137) **Liu Wenbin, J.E. Rubio** - "Optimality conditions for elliptic variational inequalities", in: "Analysis and Optimization of Systems", A. Bensoussan, J.L. Lions Eds., LNCIS 144, Springer Verlag (1990).

138) **R.E.White** - "An enthalpy formulation of the Stefan problem", SIAM J. Numer. Anal. 19(1982).

139) **R.E.White** - "A numerical solution of the entalpy formulation of the Stefan problem", SIAM J.Numer. Anal.19(1982).

140) **C.Zalinescu** - "On an abstract control problem", Numer.Funct.Anal. and Optimiz. 2(1980).

141) **J.P.Zolesio, M.C.Delfour, G.Payre** - "Optimal design of the shape of a space radiator", Proceedings of the23rd CDC, Las Vegas (1984).

142) **J.P. Zolesio, J. Sokolowski, B. Benedict** - "Shape optimization for the contact problems", in L.N.C.I.S. 59, Springer Verlag, Berlin, p. 784-799 (1984).

143) **M.Zlamal** - "On the finite element method", Numer. Math.12(1968).

144) **J.P.Yvon** - "Controle optimal de systemes gouvernes par des inequations variationnelles", Rapport de Recherche 22, IRIA, Paris (1974).

145) **K.Yosida** - "Functional analysis", Springer Verlag, Berlin (1970).